COMPUTER

THE SLOAN TECHNOLOGY SERIES

COMPUTER

A HISTORY OF THE INFORMATION MACHINE

■ SECOND EDITION ■

Martin Campbell-Kelly and William Aspray

Westview
PRESS

A Member of the Perseus Books Group

Copyright © 2004 Martin Campbell-Kelly and William Aspray

Published in the United States of America by Westview Press, A Member of the Perseus Books Group, 5500 Central Avenue, Boulder, CO 80301-2877, and in the United Kingdom by Westview Press, 12 Hid's Copse Road, Cumnor Hill, Oxford OX2 9JJ.

Find us on the world wide web at www.westviewpress.com.

Westview Press books are available at special discounts for bulk purchases in the United States by corporations, institutions, and other organizations. For more information, please contact the Special Markets Department at the Perseus Books Group, 11 Cambridge Center, Cambridge, MA 02142, or call (617) 252-5298 or (800) 255-1514, or e-mail special.markets@perseusbooks.com.

Library of Congress Cataloging-in-Publication Data
Campbell-Kelly, Martin.
    Computer : a history of the information machine / Martin Campbell-Kelly and William Aspray.— 2nd ed.
        p.   cm. — (The Sloan technology series)
    Includes bibliographical references and index.
    ISBN 0-8133-4264-3 (pbk.)
    1. Computers—History. 2. Electronic data processing—History. I. Aspray, William. II. Title. III. Series.
QA76.17.C36 2004
004'.09—dc22

                                                                    2004006325

The paper used in this publication meets the requirements of the American National Standard for Permanence of Paper for Printed Library Materials Z39.48-1984.

:: CONTENTS

THE SLOAN TECHNOLOGY SERIES

PREFACE TO THE SLOAN TECHNOLOGY SERIES

Technology is the application of science, industry, engineering, and industrial organization to create a human-built world. It has led, in developed nations, to a standard of living inconceivable a hundred years ago. The process, however, is not free of stress: by its very nature, technology brings change in society and undermines convention. It affects virtually every aspect of human endeavor: private and public institutions, economic systems, communications networks, political structures, international affiliations, the organization of societies, and the condition of human lives. The effects are not one-way; just as technology changes society, so too do societal structures, attitudes, and mores affect technology. But perhaps because technology is so rapidly and completely assimilated, the profound interplay of technology and other social endeavors in modern history has not been sufficiently recognized.

The Sloan Foundation has had a long-standing interest in deepening public understanding about modern technology, its origins, and its impact on our lives. The Sloan Technology Series, of which this volume is a part, seeks to present to the general reader the story of the development of critical technologies. The aim of the series is to convey both the technical and human dimensions of the subject: the invention and effort entailed in devising the technologies and the comforts and stresses they have introduced into contemporary life. It is hoped that the series will disclose a past that might provide perspective on the present and inform the future.

The Foundation has been guided in its development of the Sloan Technology Series by a distinguished advisory committee. We express deep gratitude to John Armstrong, Simon Michael Bessie, Samuel Y. Gibbon, Thomas P. Hughes, Victor McElheny, Robert K. Merton, Elting E. Morison (deceased), and Richard Rhodes. The Foundation has been represented on the committee by Ralph E. Gomory, Arthur L. Singer Jr., Hirsch G. Cohen, A. Frank Mayadas, and Doron Weber.

Alfred P. Sloan Foundation

:: ACKNOWLEDGMENTS

THIS BOOK has its origins in the vision of the Alfred P. Sloan Foundation that it is important for the public to understand the technology that has so profoundly reshaped Western society during the past century. When Arthur Singer of the Sloan Foundation approached us in the fall of 1991 about writing a popular history of the computer, both of us had been researching this subject for more than a decade and bemoaned the lack of any general, accessible history; but we were both busy with other projects. Without the invitation, encouragement, generous financial support, and respectful treatment we received from the Sloan Foundation, this book would never have been written.

We are grateful to a number of people and institutions for their assistance. Janet Abbate, I. Bernard Cohen, Arthur Norberg, Judy O'Neill, and Emerson Pugh—all specialists in the history of computing—shared information and offered useful suggestions. Jack Howlett and Kenneth Beauchamp shared thoughts about the 1930s that found their way into the text. Susan Rabiner gave us invaluable advice about writing for a general public, and her staff at Basic Books gave friendly and professional service with copyediting and production. Jon Agar, Steve Russ, Dorian Rutter, and John Fauvel read sample chapters to tell us how well we were achieving the tricky balance between reaching a popular audience and satisfying the academic reader—as did our students at the University of Warwick and Rutgers University. Bruce Bruemmer and Kevin Corbett at the Charles Babbage Institute, Sheldon Hochheiser at the AT&T Archives, and Andrew Goldstein and others at the IEEE Center for the History of Electrical Engineering all helped by supplying archival materials and other historical information. Jane Campbell-Kelly helped in numerous ways with the preparation of the manuscript and the bibliography. Carol Voelker helped improve the writing in the final stages and provided publicity photos. Lastly, it is a pleasure to thank Holly Hodder and Lisa Teman for carefully shepherding the second edition of our book through Westview Press. Notwithstanding these many contributions, we take full responsibility for the contents.

Martin Campbell-Kelly
William Aspray

:: INTRODUCTION TO THE SECOND EDITION

THE INTRODUCTION to the first edition notes the risk of obsolescence facing anyone who writes about the recent history of the computer, given the rapid pace of technological innovation and changing patterns of usage. Indeed, the eight-year period between the first and second editions has witnessed momentous changes in computing technology, who uses it, and what it is used for. This has been a period in which the Internet has become ascendant in Western society and the computer is increasingly used as a communication machine. The earlier traditions of the computer—as a mathematical machine and as an information machine—have not been so much supplanted as they have been supplemented by this new tradition of the computer as a communication device. These past few years have provided additional evidence of the universal character of the computer, as a machine that can be used in seemingly endless ways. What will come next is hard to predict, but that there will be a next tradition to layer on top of these earlier traditions we have no doubt.

We generally are pleased with how well the first edition weathered the changes of the past decade. The first ten chapters have required little revision. We have corrected a small number of factual errors and changed several examples to better set the stage for the final two chapters. The focus of chapter 11 has changed from explaining the new emphasis on software innovation as the personal computer matured to the theme of broadening the appeal of the computer. Sections from the original chapter 11 on the software industry, the graphical user interface, the Macintosh, and Microsoft's Windows have been shortened and revised. New sections have been added on CD-ROMs and consumer networks. Chapter 12 also has been extensively revised. We have retained sections, in slightly altered form, on the search for a world information resource, the ARPANET, and e-mail. We have greatly expanded our coverage of the World Wide Web and its predecessors, and we trace its history through the "browser wars" and the rise of the dot-com phenomenon. Lastly, we have re-vised our bibliography to reflect the many books and articles on the history of computing published over the past decade by historians, journalists, and com-puter practitioners. With these changes, we believe the book will address the in-terests of readers in the early twenty-first century.

:: INTRODUCTION TO THE
FIRST EDITION

IN JANUARY 1983, *Time* magazine selected the personal computer as its Man of the Year, and public fascination with the computer has continued to grow ever since. That year was not, however, the beginning of the computer age. Nor was it even the first time that *Time* had featured a computer on its cover. Thirty-three years earlier, in January 1950, the cover had sported an anthropomorphized image of a computer wearing a navy captain's hat to draw readers' attention to the feature story, about a calculator built at Harvard University for the U.S. Navy. Sixty years before that, in August 1890, another popular American magazine, *Scientific American,* devoted its cover to a montage of the equipment constituting the new punched-card tabulating system for processing the U.S. Census. As these magazine covers indicate, the computer has a long and rich history, and we aim to tell it in this book.

In the 1970s, when scholars began to investigate the history of computing, they were attracted to the large one-of-a-kind computers built a quarter-century earlier, sometimes now referred to as the "dinosaurs." These were the first machines to resemble in any way what we now recognize as computers: They were the first calculating systems to be readily programmed and the first to work with the lightning speed of electronics. Most of them were devoted to scientific and military applications, which meant that they were bred for their sheer number-crunching power. Searching for the prehistory of these machines, historians mapped out a line of desktop calculating machines originating in models built by the philosophers Blaise Pascal and Gottfried Leibniz in the seventeenth century and culminating in the formation of a desk calculating industry in the late nineteenth century. According to these histories, the desk calculators were followed in the period between the world wars by analog computers and electromechanical calculators for special scientific and engineering applications; the drive to improve the speed of calculating machines during World War II led directly to the modern computer.

Although correct in the main, this account is not complete. Today, research scientists and atomic weapons designers still use computers extensively, but the vast majority of computers are employed for other purposes, such as word processing and keeping business records. How did this come to pass? To answer this question, we must take a broader view of the history of the computer as the history of the information machine.

This history begins in the early nineteenth century. Because of the increasing population and urbanization in the West resulting from the Industrial Revolution, the scale of business and government grew, and with it grew the scale of information collection, processing, and communication needs. Governments began to have trouble enumerating their populations, telegraph companies could not keep pace with their message traffic, and insurance agencies had trouble processing policies for the masses of workers.

Novel and effective systems were developed for handling this increase in information. For example, the Prudential Company of England developed a highly effective system for processing insurance policies on an industrial scale using special-purpose buildings, rationalization of process, and division of labor. But by the last quarter of the century, large organizations turned increasingly to technology as the solution to their information-processing needs. On the heels of the first large American corporations came a business machine industry to supply them with typewriters, filing systems, and duplication and accounting equipment.

The desk calculator industry was part of this business machine movement. For the previous two hundred years, desk calculators had merely been hand-made curiosities for the wealthy. But by the end of the nineteenth century, these machines were being mass-produced and installed as standard office equipment, first in large corporations and later in progressively smaller offices and retail establishments. Similarly, the punched-card tabulating system developed to enable the U.S. government to cope with its 1890 census data gained wide commercial use in the first half of the twentieth century, and was in fact the origin of IBM.

Also beginning in the nineteenth century and reaching maturity in the 1920s and 1930s was a separate tradition of analog computing. Engineers built simplified physical models of their problems and measured the values they needed to calculate. Analog computers were used extensively and effectively in the design of electric power networks, dams, and aircraft.

Although the calculating technologies available through the 1930s served business and scientific users well, during World War II they were not up to the demands of the military, which wanted to break codes, prepare firing tables for new guns, and design atomic weapons. The old technologies had three shortcomings: They were too slow in doing their calculations, they required human intervention in the course of a computation, and many of the most advanced calculating systems were special-purpose rather than general-purpose devices.

Because of the exigencies of the war, the military was willing to spend whatever it would take to develop the kinds of calculating machines it needed. Millions of dollars were spent, resulting in the production of the first electronic, stored-program computers—although, ironically, none of these machines was completed in time for war work. The military and scientific research value of these computers was nevertheless appreciated, and by the time of the Korean War a small number had been built and placed in operation in military facilities, atomic energy laboratories, aerospace manufacturers, and research universities.

Although the computer had been developed for number crunching, several groups recognized its potential as a data-processing and accounting machine. The developers of the most important wartime computer, the ENIAC (Electronic Numerical Integrator and Computer), left their university posts to start a business building computers for the scientific and business markets. Other electrical manufacturers and business-machine companies, including IBM, also turned to this enterprise. The computer makers found a ready market in government agencies, insurance companies, and large manufacturers.

The basic functional specifications of the computer were set out in a government report written in 1945, and these specifications are still largely followed today. However, forty years of continuous and extraordinarily rapid-paced innovation have followed the original conception. These innovations are of two types. One is the improvement in components, leading to faster processing speed, larger information storage capacity, improved price performance, better reliability, less required maintenance, and the like—today's computers are literally millions of times better than the first computers on almost all measures of this kind. These innovations were made predominantly by the firms that manufactured computers.

The second type of innovation was in the mode of operation, but here the agent for change was most often the academic sector, backed by government financing. In most cases, these innovations became a standard part of computing only through their refinement and incorporation into standard products by the computer manufacturers. There are five notable examples of this kind of innovation: high-level programming languages, real-time computing, time-sharing, networking, and graphically oriented human-computer interfaces.

While the basic structure of the computer remained unchanged, these new components and modes of operation revolutionized our human experiences with computers. Elements that we take for granted today, such as having a computer on our own desk, equipped with a mouse, monitor, and disk drive, were not even conceivable until the 1970s. Then most computers cost hundreds of thousands, or even millions, of dollars and filled a large room. Users would seldom touch or even see the computer. Instead, they would bring a stack of punched cards representing their program to an authorized computer operator and return hours or days later to pick up a printout of their results. As the mainframe became more refined, the punched cards were replaced by remote terminals and response time from the computer became almost immediate, but still only the privileged few had access to the computer. This has all changed with the development of the personal computer and the growth of the Internet. The mainframe has not died out, as many have predicted, but computing is now available to the masses.

We have organized the book in four sections. The first covers the way computing was handled before the arrival of electronic computers. The next two sections describe the mainframe computer era, roughly from 1945 to 1980, with one section devoted to the computer's creation and the other to its evolution. The final section discusses the origins of the new computing environment of the personal computer and the Internet.

The first section, on the early history of computing, includes three chapters. Chapter 1 discusses manual information processing and early technologies. People often suppose that information processing is a twentieth-century phenomenon; this is not so, and this chapter shows that sophisticated information processing could be done with or without machines—slower in the latter case, but equally well. Chapter 2 describes the origins of office machinery and the business machine industry. To understand the modern computer industry, it is necessary to realize that its leading firms—including IBM—were established as business machine manufacturers in the last decades of the nineteenth century and were major innovators between the two world wars. Chapter 3 describes Charles Babbage's failed attempt to build a calculating engine in the 1830s and its realization by Harvard University and IBM a century later.

The second section of the book describes the development of the electronic computer, from its invention during World War II up to the establishment of IBM as the dominant mainframe computer manufacturer in the mid-1960s. Chapter 4 covers the development of the ENIAC at the University of Pennsylvania during the war, and its successor, the EDVAC, which was the blueprint for almost all subsequent computers up to the present day. Chapter 5 describes the early development of the computer industry, which transformed the computer from a scientific instrument for mathematical computation into a machine for business data processing. In chapter 6 we examine the development of the mainframe computer industry, focusing on the IBM System/360 range of computers, which created the first stable industry standard and established IBM's dominance.

The third part of the book presents a selective history of some key computer innovations in the quarter-century between the invention of the computer at the end of the war and the development of the first personal computers. Chapter 7 is a study of one of the key technologies of computing, real time. We examine this subject in the context of commonly experienced applications, such as airline reservations and supermarket bar codes. Chapter 8 describes the development of software technology and the emergence of a software industry. Chapter 9 covers the development of some of the key features of the computing environment at the end of the 1960s: time-sharing, minicomputers, and microelectronics. The purpose of the chapter is, in part, to redress the commonly held notion that the computer transformed from the mainframe to the personal computer in one giant leap.

The last section of the book gives a history of the developments of the last twenty years that brought the computer to most people's desktops. Chapter 10 describes the development of the microcomputer from the first hobby computers in the mid-1970s, up to its transformation into the familiar personal computer by the end of the decade. Chapter 11's focus is on the personal-computer environment of the 1980s, when the critical problems in personal computing were no longer hardware but software—especially "user-friendly" software. This decade was characterized by the extraordinary rise of Microsoft and the other personal-computer software companies. The book concludes with a discussion of the latest chapter in computing, the Internet. The focus is on the World Wide Web and its precedents in the information sciences.

This book, perhaps more than many others, is a product of its time. The computer technology that we are studying is evolving rapidly. Over the five years during which this book was written, we used three generations of personal computers. The material in the last section is at particular risk of looking dated within a few years, after the computing field has moved off in some new direction. The historical study of computing itself is also a young discipline. Historians have passed beyond the period where their focus was on desk calculators and number-crunching computers. Our work falls in the present generation of scholarship based on the broader definition of the information machine, with strong business and other contextual factors considered in addition to technical factors. We anticipate that within the next decade, a new body of historical scholarship will appear that will enable someone to write a new synthetic account that will deepen our understanding of computers in relation to consumers, gender, labor, and other social and cultural issues.

There are a few gaps in this book that cannot be explained by the absence of a secondary literature. We have not, for example, attempted to discuss computer science and computer theory, nor have we treated artificial intelligence. We felt that the former would be too arcane and "internalist" for the general reader, while the latter—though full of interest—has not yet advanced sufficiently for us to be able to evaluate its contributions, separate hype from reality, and incorporate the subject meaningfully into the mainstream history of computing.

We have included notes on our sources at the end of the book. These indicate the exact sources of our quotations and lead the interested reader to some of the major literature on the history of computing.

Part One

BEFORE THE COMPUTER

1

:: WHEN COMPUTERS
WERE PEOPLE

THE WORD *computer* is a misleading name for the ubiquitous machine that sits on our desks. If we go back to the Victorian period, or even the World War II era, the word meant an occupation, defined in the *Oxford English Dictionary* as "one who computes; a calculator, reckoner; *specifically* a person employed to make calculations in an observatory, in surveying, etc."

In fact, although the modern computer can work with numbers, its main use is for storing and manipulating information, that is, for doing the kinds of jobs performed by a clerk—defined in the *Oxford English Dictionary* as "one employed in a subordinate position in a public or private office, shop, warehouse, etc., to make written entries, keep accounts, make fair copies of documents, do the mechanical work of correspondence and similar 'clerkly' work."

The electronic computer can be said to combine the roles of the human computer and the human clerk.

Logarithms and Mathematical Tables

The first attempt to organize information processing on a large scale using human computers was for the production of mathematical tables, such as logarithmic and trigonometric tables. Logarithmic tables revolutionized mathematical computation in the sixteenth and seventeenth centuries by enabling time-consuming arithmetic operations, such as multiplication and division and the extraction of roots, to be performed using only the simple operations of addition and subtraction. Trigonometrical tables enabled a similar speeding up of calculations of angles and areas in connection with surveying and astronomy.

3

However, logarithmic and trigonometric tables were merely the best-known general-purpose tables. By the late eighteenth century, specialized tables were being produced for several different occupations: navigational tables for mariners, star tables for astronomers, life insurance tables for actuaries, civil engineering tables for architects, and so on. All these tables were produced by human computers, without any mechanical aid.

For a maritime nation such as Great Britain, and later the United States, the timely production of reliable navigation tables free of error was of major economic importance. In 1766 the British government sanctioned the astronomer royal, Nevil Maskelyne, to produce each year a set of navigational tables to be known as the *Nautical Almanac*. This was the first permanent table-making project to be established in the world. Often known as the Seaman's Bible, the *Nautical Almanac* dramatically improved navigational accuracy. It has been published without a break every year since 1766.

The *Nautical Almanac* was not computed directly by the Royal Observatory, but by a number of freelance human computers dotted around Great Britain. The calculations were performed twice, independently, by two computers, and checked by a third "comparator." Many of these human computers were retired clerks or clergymen with a facility for figures and a reputation for reliability who worked from home. We know almost nothing of these anonymous drudges. Probably the only one to escape oblivion was the Reverend Malachy Hitchins, an eighteenth-century Cornish clergyman who was a computer and comparator for the *Nautical Almanac* for a period of forty years. A lifetime of computational dedication earned him a place in the *Dictionary of National Biography*. When Nevil Maskelyne died in 1811—Hitchins had died two years previously—the *Nautical Almanac* "fell on evil days for about 20 years, and even became notorious for its errors."

Charles Babbage and Table Making

During this period Charles Babbage became interested in the problem of table making and the elimination of errors in tables. Born in 1791, the son of a wealthy London banker, Babbage spent his childhood in Totnes, Devon, a country town in the west of England. He experienced indifferent schooling, but succeeded in teaching himself mathematics to a considerable level. He went to Trinity College, Cambridge University, in 1810, where he studied mathematics. Cambridge was the leading English university for mathematics, and Babbage was dismayed to discover that he already knew more than his tutors. Realizing that Cambridge (and England) had become a mathematical backwater compared to continental Europe, Babbage and two fellow students organized the Analytical Society, which succeeded in making major reforms of mathematics in Cambridge and eventually the whole of England. Even as a young man, Babbage was a talented propagandist.

Babbage left Cambridge in 1814, married, and settled in Regency London to lead the life of a gentleman philosopher. His researches were mainly mathematical, and in 1816 his achievements were recognized by his election to the Royal Society, the leading scientific organization in Britain. He was then twenty-five—an enfant terrible with a growing scientific reputation.

In 1819 Babbage made the first of several visits to Paris, where he made the acquaintance of a number of the leading members of the French Scientific Academy, such as the mathematicians Pierre-Simon Laplace and Joseph Fourier, with whom he formed lasting friendships. It was probably during this visit that Babbage learned of the great French table-making project organized by Baron Gaspard de Prony. This project would show Babbage a vision that would determine the future course of his life.

De Prony began the project in 1790, shortly after the French Revolution. Napoleon had decided to reform many of France's ancient institutions, and in particular he wanted to establish a fair system of property taxation. To achieve this, he first needed up-to-date maps of France. De Prony was charged with this task and was appointed head of the Bureau du Cadastre, the French ordinance survey office. His task was made more complex because Napoleon had also decided to reform the old imperial system of weights and measures by introducing the new rational metric system. This created within the bureau the job of making a complete new set of decimal tables, to be known as the *tables du cadastre*. It was by far the largest table-making project the world had ever known, and de Prony decided to organize it much as one would organize a factory.

De Prony took as his starting point the most famous economics text of his day, Adam Smith's *Wealth of Nations,* published in 1776. It was Smith who first advocated the principle of division of labor, which he illustrated by means of a pin-making factory. In this famous example, Smith explained how the making of a pin could be divided into several distinct operations: cutting the short lengths of wire to make the pins, forming the pin head, sharpening the points, polishing the pins, packing them, and so on. If a worker specialized in a single operation, the output would be vastly greater than if a single worker performed all the operations that went into making a pin. De Prony "conceived all of a sudden the idea of applying the same method to the immense work with which I had been burdened, and to manufacture logarithms as one manufactures pins."

De Prony organized his table-making "factory" into three sections. The first section consisted of half a dozen eminent mathematicians, including Adrien Legendre and Lazare Carnot, who decided on the mathematical formulas to be used in the calculations. Beneath them was another small section—a kind of middle management—that, given the mathematical formulas to be used, organized the computations and compiled the results ready for printing. Finally, the third and largest section, which consisted of sixty to eighty human computers, did the actual computation. The computers used the "method of differences," which required only the two basic operations of addition and subtraction, and not the more demanding operations of multiplication and division. Hence the computers were not, and did not need to be, educated beyond basic numeracy and literacy. In fact, most of them were hairdressers who had lost their jobs because "one of the most hated symbols of the ancient régime was the hairstyles of the aristocracy."

Although the Bureau was producing mathematical tables, the operation was not itself mathematical. It was fundamentally the application of an organizational technology, probably for the first time outside a manufacturing or military context, to the production of information. Its like would not be seen again for another forty years.

The whole project lasted about a decade, and by 1801 the tables existed in manuscript form all ready for printing. Unfortunately, for the next several decades, France was wracked by one financial and political crisis after another, so that the large sum of money needed to print the tables was never found. Hence, when Babbage learned of the project in 1819, all there was to show of it was the manuscript tables in the library of the French Scientific Academy.

In 1820, back in England, Babbage gained some firsthand experience of table making while preparing a set of star tables for the Astronomical Society, a scientific society that he and a group of like-minded amateur scientists had established the same year. Babbage and his friend John Herschel were supervising the construction of the star tables, which were being computed in the manner of the *Nautical Almanac* by freelance computers. Babbage's and Herschel's roles were to check the accuracy of the calculations and to supervise the compilation and printing of the results. Babbage complained of the difficulty of table making, finding it error-prone and tedious; and if he found it tedious just supervising the table making, so much the worse for those who did the actual computing.

Babbage's unique role in nineteenth-century information processing was due to the fact that he was in equal measure a mathematician and an economist. The mathematician in him recognized the need for reliable tables and knew how to make them, but it was the economist in him that saw the significance of de Prony's organizational technology and had the ability to carry the idea further.

De Prony had devised his table-making operation using the principles of mass production at a time when factory organization involved manual labor using very simple tools. But in the thirty years since de Prony's project, best practice in factories had itself moved on, and a new age of mass-production machinery was beginning to dawn. The laborers in Adam Smith's pin-making factory would soon be replaced by a pin-making machine. Babbage decided that rather than emulate de Prony's labor-intensive and expensive manual table-making organization, he would ride the wave of the emerging mass-production technology and invent a machine for making tables.

Babbage called his machine a Difference Engine because it would use the same method of differences that de Prony and others used in table making. Babbage knew, however, that most errors in tables came not from calculating them but from printing them, so he designed his engine to set the type ready for printing as well. Conceptually the Difference Engine was very simple: It consisted of a set of adding mechanisms to do the calculations and a printing part.

Babbage applied his considerable skills as a publicist to promote the idea of the Difference Engine. He began his campaign by writing an open letter to the president of the Royal Society, Sir Humphrey Davy, in 1822, proposing that the government finance him to build the engine. Babbage argued that high-quality tables were essential for a maritime and industrial nation, and that his Difference Engine would be far cheaper than the nearly one hundred overseers and human computers in de Prony's table-making project. He had the letter printed at his own expense and ensured that it got into the hands of people of influence. As a result in 1823 he obtained government funding of £1,500 to build the Difference Engine, with the understanding that more money would be provided if necessary.

Babbage managed to rally much of the scientific community to support his project. His boosters invariably argued that the merit of his Difference Engine was that it would eliminate the possibility of errors in tables "through the unerring certainty of mechanism." It was also darkly hinted that the errors in the *Nautical Almanac* and other tables might "render the navigator liable to be led into difficulties, if not danger." Babbage's friend John Herschel went further and wrote: "An undetected error in a logarithmic table is like a sunken rock at sea yet undiscovered, upon which it is impossible to say what wrecks may have taken place." Gradually the danger of errors in tables grew into lurid tales that "navigational tables were full of errors which continually led to ships being wrecked." Historians have found no evidence for this claim, although reliable tables certainly helped Britain's maritime activity run smoothly.

Unfortunately, the engineering was more complicated than the conceptualization. Babbage completely underestimated the financial and technical resources he would need to build his engine. He was at the cutting edge of production technology, for although relatively crude machines such as steam engines and power looms were in widespread use, sophisticated devices such as pin-making machines were still a novelty. By the 1850s such machinery would be commonplace, and there would exist a mechanical-engineering infrastructure that made building them relatively easy. While building the Difference Engine in the 1820s was not in any sense impossible, Babbage was paying the price of being a first mover; it was rather like building the first computers in the mid-1940s: difficult and extremely expensive.

Babbage was now battling on two fronts: first, designing the Difference Engine; and second, developing the technology to build it. Although the Difference Engine was conceptually simple, its design was mechanically complex. In the London Science Museum, today, one can see evidence of this complexity in hundreds of Babbage's machine drawings for the engines, and thousands of pages of his notebooks. During the 1820s Babbage scoured the factories of Europe seeking gadgets and technology that he could use in the Difference Engine. Not many of his discoveries found their way into the Difference Engine, but he succeeded in turning himself into the most knowledgeable economist of manufacturing of his day. In 1832 he published his most important book, an economics classic titled *Economy of Manufactures*, which ran to four editions and was translated into five languages. In the history of economics, Babbage is a seminal figure who connects Adam Smith's *Wealth of Nations* to the Scientific Management movement, founded in America by Frederick Winslow Taylor in the 1880s.

The government continued to advance Babbage money during the 1820s and early 1830s, eventually totaling £17,000; and Babbage claimed to have spent much the same again from his own pocket. These were very large sums in today's money. By 1833, Babbage had produced a beautifully engineered prototype Difference Engine that was too small for real table making and lacked a printing unit, but showed beyond any question the feasibility of his concept. (It is still on permanent exhibit in London's Science Museum, and it works as perfectly today as it did then.)

To develop a full-scale machine Babbage needed even more money, which he requested in a letter in 1834 to the prime minister, the Duke of Wellington. Unfortunately, at that time, Babbage had an idea of such stunning originality that he just could not keep quiet about it: a new kind of engine that would do all the Difference Engine could do but much more—it would be capable of performing *any* calculation that a human could specify for it. This machine he called the Analytical Engine. In almost all important respects, it had the same logical organization as the modern electronic computer. In his letter to the Duke of Wellington, Babbage hinted that instead of completing the Difference Engine he should be allowed to build the Analytical Engine. Raising the specter of the Analytical Engine was the most spectacular political misjudgment of Babbage's career; it fatally undermined the government's confidence in his project, and he never obtained another penny. In fact, by this time, Babbage had become so thoroughly immersed in his calculating engine project that he had completely lost sight of the original objective: to make tables. The engines had become an end in themselves, as we shall see in chapter 3.

Clearing Houses and Telegraphs

While Babbage was struggling with his Difference Engine, the idea of large-scale information processing was highly unusual—whether it was organized manually or using machinery. The volume of activity in ordinary offices of the 1820s simply did not call for large clerical staffs. Nor was there any office machinery to be had; even adding machines were little more than a scientific novelty at this date, and the typewriter had yet to be invented. For example, the Equitable Society of London—then the largest life insurance office in the world—was entirely managed by an office staff of eight clerks, equipped with nothing more than quill pens and writing paper.

In the whole of England there was just one large-scale data-processing organization that had an organizational technology comparable with de Prony's table-making project. This was the Bankers' Clearing House in the City of London, and Babbage wrote the only contemporary published account of it.

The Bankers' Clearing House was an organization that processed the rapidly increasing number of checks being used in commerce. When the use of checks became popular in the eighteenth century, a bank clerk physically had to take a check deposited by a customer to the bank that issued it to have it exchanged for cash. As the use of checks gained in popularity in the middle of the eighteenth century, each of the London banks employed a "walk clerk," whose function was to make a tour of all the other banks in the City, the financial district of London, exchanging checks for cash. In the 1770s, this arrangement was simplified by having all the clerks meet at the same time in the Five Bells Public House on Lombard Street. There they performed all the exchanging of checks and cash in one "clearing room." This obviously saved a lot of walking time and avoided the danger of robbery. It also brought to light that if two banks had checks drawn on each other, the amount of cash needed for settlement was simply the difference between the two amounts owed, which was usually far less than the total amount of all the checks. As the volume of business expanded, the clearing room outgrew its premises and moved several times. Eventually, in the early 1830s, the London banks jointly built a Bankers' Clearing House at 10 Lombard Street, in the heart of London's financial center.

The Bankers' Clearing House was a secretive organization that shunned visitors and publicity. This was because the established banks wanted to exclude the many banks newly formed in the 1820s (which the Clearing House succeeded in doing until the 1850s). Babbage, however, was fascinated by the clearing house concept and pulled strings to gain entry. The secretary of the Bankers' Clearing House was a remarkable man by the name of John Lubbock, who, besides being a leading figure in the City, was also an influential amateur scientist and vice president of the Royal Society. Babbage wrote to Lubbock in October 1832 asking if it were "possible that a stranger be permitted as a spectator." Lubbock replied, "*You* can be taken to the clearing house . . . but we wish it not to be mentioned, so that the public may fancy they can have access to the sanctum sanctorum of banking, and we wish of course not to be named."

Babbage was captivated by Lubbock's scientifically organized system which, despite Lubbock's proscription, he described in glowing terms in the *Economy of Manufactures*:

> In a large room in Lombard Street, about thirty clerks from the several London bankers take their stations, in alphabetical order, at desks placed round the room; each having a small open box by his side, and the name of the firm to which he belongs in large characters on the wall above his head. From time to time other clerks from every house enter the room, and, passing along, drop into the box the checks due by that firm to the house from which this distributor is sent.

Most of the day was spent by clerks exchanging checks with one another and entering the details in ledgers. At four o'clock in the afternoon, the settlements between the banks would begin. The clerk for each bank would total all the checks received from other banks for payment and all the checks presented for payment to other banks. The difference between these two numbers would then be either paid out or collected.

At five o'clock the inspector of the clearing house took his seat on a rostrum in the center of the room. Then, the clerks from all the banks who owed money on that day paid the amount due in cash to the inspector. Next, the clerks of all the banks that were owed money collected their cash from the inspector. Assuming that no mistakes had been made (and an elaborate accounting system using preprinted forms ensured this did not happen very often), the inspector was left with a cash balance of exactly zero.

The amount of money flowing through the Bankers' Clearing House was staggering. In the year 1839, £954 million was cleared—the equivalent of several hundred billion dollars in today's money. On the busiest single day over £6 million was cleared, and about half a million pounds in bank notes were used for the settlement. Eventually, the need for cash was eliminated altogether by each bank and the Clearing House having an account with the Bank of England. Settlements were then made simply by transferring an amount from a bank's account to that of the Clearing House, or vice versa.

Babbage clearly recognized the significance of the Bankers' Clearing House as an example of the "division of mental labor," comparable with de Prony's table-making project and his own Difference Engine. He remained deeply interested in large-scale information processing all his life. For example, in 1837 he applied unsuccessfully to become registrar general and director of the English population census. But by the time really big information-processing organizations came along in the 1850s and 1860s—such as the great savings banks and industrial assurance companies—Babbage was an aging man who had ceased to have any influence.

The Bankers' Clearing House was an early example of what would today be called financial infrastructure. The Victorian period was the great age of physical and financial infrastructure investment. Between 1840 and 1870, Britain's investment in rail track grew from 1,500 to over 13,000 miles. Alongside this physical and highly visible transport infrastructure grew a parallel, unseen information infrastructure known as the Railway Clearing House, which was modeled very closely on the Bankers' Clearing House. Established in 1842, the Railway Clearing House rapidly became one of the largest data-processing bureaucracies in the world. By 1870 there were over 1,300 clerks processing nearly 5 million transactions a year.

Another vital part of the Victorian information infrastructure was the telegraph, which began to compete with the ordinary postal system in the 1860s. Telegrams were expensive—one shilling for a twenty-word message compared with as much as you could want to write in a letter for one penny—but very fast. A telegram would speed across the country in an instant and arrive at its final destination in as little as an hour, hand-delivered by a bicycling telegraph boy.

Rather like the Internet in our own time, the telegraph did not originate as a planned communications system. Instead it began as a solution to a communications problem in the early rail system. There was a widely held fear among the public of a passenger train entering a section of track while another was heading in the opposite direction (in practice there were very few such accidents, but this did not diminish public concern). To solve the problem inventive engineers strung up an electrical communication system alongside the track so that the signalmen at each end could communicate. Now a train could not enter a single-track section until two human operators agreed that it was safe to do so. Of course, it was not long before a commercial use was found for the new electrical signaling method. Newspapers and commercial organizations were willing to pay for news and market information ahead of their competitors. Suddenly, telegraph poles sprang up alongside railway tracks—some systems were owned by the railway companies, some by newly formed telegraph companies. Although messages were sent by electricity, the telegraph still needed a large clerical labor force to operate the machines that transmitted the messages. Much of the labor was female—the first time that female clerical labor had been used on any scale in Britain. The reason for this was that telegraph instruments were rather delicate and it was believed that women—especially seamstresses familiar with sewing machines—would have more dexterity than men.

By the mid-1860s there were over 75,000 miles of telegraph lines in Britain, operated by six main companies. However, each system operated independently and it was difficult for a telegram originating in one network to make use of another. In 1870, the British government stepped in to integrate the systems into a national telegraph network. Once this was done, telegraph usage simply exploded. More telegraph lines were erected and old ones were renewed. Telegraph offices were established in every significant town. Telegraph schools were set up in London and the provinces to train young men and women in Morse telegraphy. The cost of a telegram fell to sixpence for a dozen words.

The transmission of telegrams presented some interesting technical problems. Chief among these was the need to send telegrams between locations that were not directly connected by a telegraph line. Consider the problem of a cigar manufacturer in Edinburgh, Scotland, negotiating with a tobacco importer in Bristol, England, some 350 miles south. There was no direct connection between these two great cities. Instead the telegram had to be passed—rather like a baton in a relay race—through the telegraph offices of intermediate cities. Edinburgh to Newcastle, Newcastle to York, York to Manchester, Manchester to Birmingham, and finally Birmingham to Bristol. At each intervening telegraph office the message was received by a telegraphist on a Morse sounder and written out in longhand. The message would then be resent by another telegraphist to the next telegraph office in the chain. Although labor-intensive, the system was very resilient. If the telegraph lines between York and Manchester, say, were storm-damaged or simply very busy the operator might send the message via Sheffield, which was not too much of a diversion. Sheffield would then send the message on its southerly route. Telegraphists needed to have a good knowledge of national geography.

After the government took over the telegraph system, and because London was the political and commercial center of Britain, it made sense for all major cities to have direct lines into the capital. In 1874 a central hub, the Central Telegraph Office, was established with a direct connection to "every town of importance in the United Kingdom." The Central Telegraph Office occupied a purpose-built structure in St. Martin's Le Grand, sited between Parliament, and the financial district and the newspaper offices in Fleet Street. The office was the epitome of scientific modernity and featured in illustrated books and magazines. From the day it began operations, the great majority of national telegraph traffic passed through it—now our cigar manufacturer in Edinburgh could communicate with his tobacco importer in Bristol with just a single hop—this was faster, cheaper, and less likely to introduce transcription errors in the message.

In 1874, the *Illustrated London News*—a magazine inclined to technological boosterism—produced a full-page engraving of a gallery in the Central Telegraph Office. The image shows an information factory frozen in time—row upon row of young men and women are operating telegraph instruments, supervisors (generally just a little older than their charges) organize the work from a massive sorting table at the front of the room, while messenger boys (mostly fresh out of elementary school) run up and down the rows of telegraph stations collecting messages as they are transcribed and distributing them for onward transmission. The writer of the article explained—in words that spoke not only of the telegraph, but of the times in which he lived:

> It is a cheerful scene of orderly industry, and it is, of course, not the less pleasing because the majority of the persons here are young women, looking brisk and happy, not to say pretty, and certainly quite at home. Each has her own instrument on the desk before her. She is either just now actually busied in working off or in reading some message, or else, for the moment she awaits the signal, from a distant station, to announce a message for her reception. Boys move here and there about the galleries, with the forms of telegrams, which have been received in one part of the instrument-room, and which have to be signaled from another, but which have first to be conveyed, for record, to the nearest check-tables and sorting-tables in the centre.

The journalist—evidently a man with a taste for statistics—noted that there were 1,200 telegraphists, of whom 740 were female, and 270 boy messengers. Each day between 17,000 and 18,000 messages were transmitted between provincial telegraph offices, while nearly as many again were transmitted within London. But this was only the beginning. By the turn of the century, the Central Telegraph Office employed no less than 4,500 clerks and transmitted between 120,000 and 165,000 telegrams a day. It was the largest office in the world.

Herman Hollerith and the 1890 Census

Compared to Europe, the United States was a latecomer to large-scale data processing. That's because it lagged twenty or thirty years behind Europe in its economic development. When Britain, Germany, and France were industrializing in the 1830s, the United States was still primarily an agricultural country. It was not until after the Civil War that U.S. businesses began to develop big offices, but this delay enabled them to take full advantage of the newly emerging office technologies.

Before the Civil War the only American data-processing bureaucracy of major importance was the Bureau of the Census in Washington, D.C. The population census was established by an act of Congress in 1790 to determine the "apportionment" of members of the House of Representatives. The first census in 1790 estimated the population of the United States as 3.9 million, and consequently established that there should be an apportionment of one representative for each 33,000 people, or 105 representatives. The early census data processing was very small scale, and no records exist of how it was organized. Even by 1840, when the population was 17.1 million, there were still only 28 clerks in the Bureau of the Census. Twenty years later, however, by the 1860 census, there was a major bureaucracy that employed 184 clerks to count a population of 31.4 million. For the 1870 census, there were 438 clerks, and the census reports amounted to 3,473 pages.

After that, the growth of the census was simply explosive. The census of 1880 probably represented a high point in manual data processing in the United States, when there were no fewer than 1,495 clerks employed to process the census data. The data-processing method used was known as the "tally system." This can best be understood by an example. One of the reports produced by the census was a table of the age structure of the population for each of the states, and the major cities and towns (that is, the number of males and females, in their ethnic categories, of each age). A tally clerk was provided with a large tally sheet, ruled into a grid, with a dozen columns and many rows. Each pair of columns corresponded to males and females of a particular ethnic group. The rows corresponded to the age of an individual—under one year of age, one year of age, two years of age, and so on, up to a hundred years of age and over. The tally clerk would take a stack of census forms from an "enumeration district" (representing about a hundred families), and would proceed to examine the age, sex, and ethnic origin of each person on each form, putting a check mark in the appropriate cell of the grid. When the pile of forms had been processed in this way, the clerk would count the number of check marks in each of the cells and write the result at the side in red ink. This would be repeated for every enumeration district in the city. Finally, another clerk would total all the red-inked numbers on all the tally sheets for a city, entering the results onto a consolidation sheet, which would eventually become one of the tables in the census report.

The work of the census clerks was tedious beyond belief—so much so that a journalist of the period wrote, "The only wonder . . . is, that many of the clerks who toiled at the irritating slips of tally paper in the census of 1880 did not go blind and crazy." More than 21,000 pages of census reports were produced for the 1880 census, which took some seven years to process. This unreasonably long time provided a strong motive to speed up the census by mechanization or any other means that could be devised.

One person who was acutely aware of the census problem was a remarkable young engineer by the name of Herman Hollerith (1859–1929). He later developed a mechanical system for census data processing, commercialized his invention by establishing the Tabulating Machine Company in 1896, and laid the foundations of IBM. Along with Babbage, Hollerith is regarded as one of the seminal nineteenth-century figures in the development of information processing. Though Hollerith was not a deep thinker like the polymath Babbage, he was practical where Babbage was not. Hollerith also had a strong entrepreneurial flair, so that he was able to exploit his inventions and establish a major industry.

Hollerith grew up in New York and attended Columbia University, where one of his professors was an adviser to the Bureau of the Census in Washington. He invited the newly graduated Hollerith to become his assistant. While with the Census Bureau, Hollerith saw for himself the extraordinary scale of clerical activity—there was nothing else in the country to compare with it at the time. This familiarity with census operations enabled him to devise an electrical tabulating system that would mechanize much of the clerical drudgery.

Hollerith's key idea was to record the census return for each individual as a pattern of holes on punched paper tape or a set of punched cards, similar to the way music was recorded on a string of punched cards on fairground organettes of the period. It would then be possible to use a machine to automatically count the holes and produce the tabulations. In later years Hollerith would reminisce on the origins of the idea: "I was traveling in the West and I had a ticket with what I think was called a punch photograph. [The conductor] punched out a description of the individual, as light hair, dark eyes, large nose, etc. So you see, I only made a punch photograph of each person."

In 1888 the preparations for the 1890 census had begun under the direction of a newly appointed superintendent of the census, Robert P. Porter. Porter, an English-born naturalized American, was a charismatic individual. He was a diplomat, an economist, and a journalist; founder and editor of the *New York Press*; an industrial pundit; and a well-known exponent on statistics. As soon as he was appointed superintendent, he set up a commission to select by competition an alternative to the system of tally sheets used in the 1880 and previous censuses. He was already an enthusiast for Hollerith's system; as a journalist he had written an article about it, and would later become chairman of the British Tabulating Machine Company (the ancestor of International Computers, which became Europe's largest computer company). But to ensure a fair contest, he disallowed himself from being a judge of the competition.

Three inventors entered the competition, including Hollerith, and all of them proposed using cards or slips of paper in place of the old tally sheets. One of Hollerith's rivals proposed transcribing the census return for each individual onto a slip of paper, using different colored inks for the different questions, so that the data could be easily identified and rapidly counted and sorted by hand.

The second competitor had much the same idea, but used ordinary ink and different colored cards, which would then be easy to identify and arrange by hand. These two systems were entirely manual and were similar to the card-based record systems that were beginning to emerge ad hoc in big commercial offices. The great advantage of the Hollerith system over these others was that once the cards had been punched, all the sorting and counting would be handled mechanically.

In the fall of 1889, the three competitors were required to demonstrate their systems by processing the 10,491 returns of the 1880 population census for the district of St. Louis. The trial involved both recording the returns on the cards or paper slips and tabulating them to produce the required statistical tables. So far as recording the data on cards was concerned, the Hollerith system proved little faster than the competing manual systems. But in the tabulation phase it came into its own, proving up to ten times as fast as the rival systems. Moreover, once the cards had been punched, the more tabulations that were required the better, the Hollerith system would prove. The commission was unanimous in recommending that the Hollerith Electric Tabulating System be adopted for the 1890 census.

As preparations for the eleventh United States population census got into high gear, Superintendent Porter "rounded up what appeared to be every square foot of empty office and loft space in downtown Washington." At the same time Hollerith made the transition from inventor to supervising manufacturer and subcontracted Western Electric to assemble his census machines. He also negotiated with paper manufacturers to supply the 60 million-plus manila cards that would be needed for the census.

On 1 June 1890, an army of 45,000 census enumerators swept the nation, collecting and checking the completed schedules from the 13 million households and dispatching them to Washington. At the Census Bureau two thousand clerks were assembled in readiness, and on 1 July they began to process the greatest and most complete census the world had ever seen. It was a moment for the American people to "feel their nation's muscle."

On 16 August 1890, six weeks into the processing of the census, the grand total of 62,622,250 was announced. But this was not what the allegedly fastest-growing nation in the world wanted to hear:

> The statement by Mr. Porter that the population of this great republic was only 62,622,250 sent into spasms of indignation a great many people who had made up their minds that the dignity of the republic could only be supported on a total of 75,000,000. Hence there was a howl, not of "deep-mouthed welcome,"

but of frantic disappointment. And then the publication of the figures of New York! Rachel weeping for her lost children and refusing to be comforted was a mere puppet-show compared with some of our New York politicians over the strayed and stolen of Manhattan Island citizens.

The press loved the story. In an article headlined "Useless Machines" the *Boston Herald* roasted Porter and Hollerith; "Slip Shod Work Has Spoiled the Census," exclaimed the *New York Herald*; and the other papers soon took up the story. But Hollerith and Porter never had any real doubts about the system.

After the rough count was over and the initial flurry of public interest had subsided, the Census Bureau settled into a routine. Recording all the data onto the cards was a task that kept the seven hundred card punches in almost constant operation. A punching clerk—doing what was optimistically described as "vastly interesting" work—could punch an average of seven hundred cards in a six-and-a-half-hour day. Female labor was heavily used for the first time in the census, which a male journalist noted "augurs well for its conscientious performance" because "women show a moral sense of responsibility that is still beyond the average." More than 62 million cards were punched, one for each citizen.

The cards then were processed by the census machines, each of which could do the work of twenty of the old tally clerks. Even so, the original fifty-six census machines eventually had to be increased to a hundred (and the extra rentals significantly added to Hollerith's income). Each census machine consisted of two parts: a tabulating machine, which could count the holes in a batch of cards, and the sorting box, into which cards were placed by the operator ready for the next tabulating operation. A census machine operator processed the cards one by one, by reading the information using a "press" that consisted of a plate of 288 retractable spring-loaded pins. When the press was forced down on the card, a pin meeting the solid material was pushed back into the press and had no effect. But a pin encountering a hole passed straight through, dipped into a mercury cup, and completed an electrical circuit. This circuit would then be used to add unity to one of forty counters on the front of the census machine. The circuit could also cause the lid of one of the twenty-four compartments of the sorting box to fly open—into which the operator would place the card so that it would be ready for the next phase of the tabulation.

Thus, if one counter of the tabulating machine and one sorting box compartment were wired up for a male subject and another counter and compartment were wired up for a female subject, then by reading a batch of cards through the machine it would be possible to determine the number of males and females represented, and to separate the cards accordingly. In practice, counts were much more complicated than this, in order to extract as much information as possible from the cards. Counts were designed to use all forty counters on the census machine and as many as possible of the twenty-four compartments in the sorting box.

At any one time there would be upward of eighty clerks operating the machines. Each operator processed at least a thousand cards an hour. In a typical day the combined operation would eat through nearly half a million cards: "In other words, the force was piercing its way through the mass at the rate of 500 feet daily, and handling a stack of cards nearly as high as the Washington Monument." As each card was read, the census machine gave a ring of a bell to indicate that it had been correctly sensed. The floor full of machines ringing away made a sound "for all the world like that of sleighing," commented a journalist. It was an awe-inspiring sight and sound; and, as the same journalist noted, "the apparatus works as unerringly as the mills of the Gods, but beats them hollow as to speed."

Hollerith personally supervised the whole operation, combating both natural and unnatural mechanical breakdown. One veteran of the census recounted:

> Herman Hollerith used to visit the premises frequently. I remember him as very tall and dark. Mechanics were there frequently too, to get ailing machines back in operation and loafing employees back at work. The trouble was usually that somebody had extracted the mercury from one of the little cups with an eye-dropper and squirted it into a spittoon, just to get some unneeded rest.

The Census Bureau used the Hollerith system not only to reduce the cost of the census, but also to improve the quality and quantity of information, and the speed with which it was produced. Thus the 1890 census was processed in two and a half years, compared with the seven years of the previous census. All together, the census reports totaled 26,408 pages. The total cost was $11.5 million, and it was estimated that without the Hollerith system it would have cost $5 million more. For those with eyes to see, the Hollerith machines had opened up a whole new vista of mechanized information processing.

America's Love Affair with Office Machinery

While the Hollerith system was the most visible use of mechanical information processing in the United States, it was really only one of many examples of what we now call "information technology" that developed in the twenty years following the Civil War. The Hollerith system was not in any sense typical of this information technology. Most office machines were much more humdrum: typewriters, record-keeping systems, and adding machines. Even more humdrum were the many varieties of office supplies available to American business: a hundred different types and grades of pencil; dozens of makes of fountain pens; paper clips, fasteners, and staplers of every conceivable variety; patented check-cutting machines to prevent fraud; gadgets for sorting cash, and cashier's coin trays; carbon paper and typewriter sundries; loose-leaf ledgers and filing cabinets; wooden desks equipped with pigeonholes. The list goes on and on.

In the closing decades of the nineteenth century, office equipment, in both its most advanced and its least sophisticated forms, was almost entirely an American phenomenon. Its like would not be seen in Europe until the new century, and in many businesses not until after World War I.

There were two main reasons for America's affinity for office machinery. First, because of the American office's late start compared with Europe, it did not carry the albatross of old-fashioned offices and entrenched archaic working methods. For example, back in England, the British Prudential Company had been established in the 1850s with Victorian data-processing methods that it could not shake off because it was never cost-effective to reengineer the office system to make use of typewriters, adding machines, or modern card index systems. Indeed, there was not a single typewriter in the Prudential before the turn of the century, and it was not until 1915 that any advanced office machinery was introduced. By contrast, the American Prudential Company in New York, which was established twenty years after the British company, immediately made use of any office appliances on the market and became a recognized leader in using office technology—a reputation it sustained right into the 1950s, when it was one of the first American offices to computerize. Similarly, while the British clearing houses were unmechanized until well into the new century, their American counterparts became massive users of Comptometers and Burroughs adding machines during the 1890s.

But no simple economic explanation can account fully for America's love affair with office machinery. The fact is that America was gadget-happy and was caught up by the glamour of the mechanical office. American firms often bought office appliances simply because they were perceived as modern—much as American firms bought the first computers in the 1950s. This attitude was reinforced by the rhetoric of the office systems movement.

Just as Frederick W. Taylor was pioneering scientific management in American industry in the 1880s, focusing on the shop floor, a new breed of scientific

manager—or "systematizer"—was beginning to revolutionize the American office. As an early systematizer puffed to his audience in 1886:

> Now, administration without records is like music without notes—by ear.
> Good as far as it goes which is but a little way—it bequeathes nothing to the
> future. . . . Under rational management the accumulation of experience, and
> its systematic use and application, form the first fighting line.

Systematizers set about restructuring the office, introducing typewriters and
adding machines, designing multipart business forms and loose-leaf filing
systems, replacing old-fashioned accounting ledgers with machine billing systems, and so on. The office systematizer was the ancestor of today's information-
technology consultant.

Powered by the fad for office rationalization, the United States was the first
country in the world to adopt office machinery on a large scale. This early start
enabled the United States to become the leading producer of information-
technology goods, a position it has sustained to the present day. In turn, the
United States dominated the typewriter, record-keeping, and adding machine
industries for most of their histories; it dominated the accounting machine industry between the two world wars; it established the computer industry after
World War II; and it dominates the personal-computer industry today. There is
thus an unbroken line of descent from the giant office-machine firms of the
1890s to the computer makers of today.

2

:: THE MECHANICAL OFFICE

IN 1928 the world's top four office-machine suppliers were Remington Rand, with annual sales of $60 million; National Cash Register (NCR), with sales of $50 million; the Burroughs Adding Machine Company, with $32 million worth of business; and—trailing the three leaders by a considerable margin—IBM, with an income of $20 million. Forty years later those same firms were among the top ten computer manufacturers, and of the four, IBM, whose sales exceeded those of the other three combined, was the third-largest corporation in the world, with annual revenues of $21 billion and a workforce of a third of a million people.

To understand the development of the computer industry, and how this apparently new industry was shaped by the past, one must understand the rise of the office-machine giants in the years around the turn of the twentieth century. This understanding is necessary, above all, to appreciate how IBM's managerial style, sales ethos, and technologies combined to make it perfectly adapted to shape and then dominate the computer industry.

Today, we use computers in the office for three main tasks. There is document preparation: for example, using a word-processing program to produce letters and reports. There is information storage: for example, using a database program to store names and addresses, or inventories. And there is financial analysis and accounting: for example, using a spreadsheet program for financial forecasting, or using a computer to organize a company's payroll.

These were precisely the three key office activities that the business-machine companies of the late nineteenth century were established to serve. Remington Rand was the leading supplier of typewriters for document preparation and filing systems for information storage. Burroughs dominated the market for adding machines used for simple calculations. IBM dominated the market for punched-card accounting machines. And NCR, having started out making cash registers in the 1880s, also became a major supplier of accounting machines.

The Typewriter

The present is shaped by the past. Nowhere is this more evident than in the history of the typewriter. Little more than a century ago the female office worker was a rarity; and office jobs were held almost exclusively by men. Today, so much office work is done by women that this often goes unremarked. It was above all the opportunities created by the typewriter that ushered female workers into the office.

A much smaller example of the past shaping the present can be seen on the top row of letters on a present-day computer keyboard: QWERTYUIOP. This wonderfully inconvenient arrangement was used in the first commercially successful typewriter produced by Remington in 1874. Because of the investment in the training of typists, and the natural reluctance of QWERTY-familiar people to change, there has never been a right moment to change to a more convenient keyboard layout. And maybe there never will be.

Prior to the development of the inexpensive, reliable typewriter, one of the most common office occupations was that of a "writer" or "copyist." These were clerks who wrote out documents in longhand. There were many attempts at inventing typewriters in the middle decades of the nineteenth century, but none of them was commercially successful because none could overcome both of the critical problems of document preparation: the difficulty of reading handwritten documents and the time it took a clerk to write them. In the nineteenth century virtually all business documents were handwritten, and the busy executive spent countless hours deciphering them. Hence the major attraction of the typewriter was that typewritten documents could be read effortlessly at several times the speed of handwritten ones. Unfortunately, using these early typewriters was a very slow business. Until the 1870s no machine came even close to the twenty-five-words-per-minute handwriting speed of an ordinary copyist. Not until typewriter speeds could rival those of handwriting did the typewriter catch on.

Newfangled typewriting machines were scientific novelties, and articles describing them occasionally appeared in technical magazines. As a result of reading such an article in the *Scientific American* in 1867, Christopher Latham Sholes, a retired newspaper editor, was inspired to invent and take out a series of patents on a new style of typewriter. This was the machine that became the first Remington typewriter. What differentiated Sholes's invention from its predecessors was its speed of operation, which was far greater than any other machine invented up to that date. This speed was due to a "keyboard and type-basket" arrangement, variations of which became almost universal in manual typewriters.

Sholes needed money to develop his invention, though, so he wrote—or rather typed—letters to all the financiers he knew of, offering to "let them in on the ground floor in return for ready funds." One of the letters was received by a Pennsylvania business promoter named James Densmore. A former newspaperman and printer, Densmore immediately recognized the significance of Sholes's invention and offered him the necessary backing. Over the next three or four years, Densmore's money enabled Sholes to perfect his machine and ready it for the marketplace.

One of the problems of the early models was that when operated at high speed, the type-bars would clash and jam the machine. On the very first machines, the letters of the keyboard had been arranged in alphabetical order, and the major cause of the jamming was the proximity of commonly occurring letter pairs (such as D and E, or S and T). The easiest way around this jamming problem was to arrange the letters in the type-basket so that they were less likely to collide. The result was the QWERTY keyboard layout that is still with us. (Incidentally, a vestige of the original alphabetical ordering can be seen on the middle row of the keyboard, where the sequence FGHJKL appears.)

Densmore made two attempts to get the Sholes typewriter manufactured by small engineering workshops, but they both lacked the necessary capital and skill to manufacture successfully and cheaply. As one historian of manufacturing has noted, the "typewriter was the most complex mechanism mass produced by American industry, public or private, in the nineteenth century." Not only did the typewriter contain hundreds of moving parts, it also used unfamiliar new materials such as rubber, sheet iron, glass, and steel.

In 1873 Densmore managed to interest Philo Remington in manufacturing the typewriter. Remington was the proprietor of a small-arms manufacturer, E. Remington and Sons, of Ilion, New York. He had no great interest in office machinery, but in the years after the Civil War the firm, needing to turn swords into plowshares, started to manufacture items such as sewing machines, agricultural equipment, and fire engines. Remington agreed to make a thousand typewriters.

The first ones were sold in 1874. Sales were slow at first because there was no ready market for the typewriter. Business offices had not yet begun to use machines of any kind, so the first customers were individuals such as "reporters, lawyers, editors, authors, and clergymen." One early enthusiast, for example, was Samuel Langhorne Clemens—better known as Mark Twain—who wrote for Remington one of the most famous testimonials ever:

> Gentlemen:
>
> Please do not use my name in any way. Please do not even divulge the fact that I own a machine. I have entirely stopped using the Type-Writer, for the reason that I never could write a letter with it to anybody without receiving a request by return mail that I would not only describe the machine but state what progress I had made in the use of it, etc., etc. I don't like to write letters, and so I don't want people to know that I own this curiosity breeding little joker.
>
> Yours truly,
Saml. L. Clemens.

It took five years for Remington to sell its first thousand machines. During this period the typewriter was further perfected. In 1878 the company introduced the Remington Number 2 typewriter, which had a shift key that raised and lowered the carriage so that both upper and lowercase letters could be printed. (The first Remington typewriter printed only capital letters.)

By 1880 Remington was making more than a thousand machines a year and had a virtual monopoly on the typewriter business. One of the key business problems that Remington addressed at an early date was sales and after-sales service. These typewriters were delicate instruments; they were prone to break down and required trained repair personnel. In this respect, the typewriter was a little like the sewing machine, which had become a mass-produced item about a decade earlier. Following Singer's lead, the Remington Typewriter Company began to establish local branches in the major cities from which typewriters were sold and to which they could be returned for repair. Remington tried to get Singer to take over its overseas business through its branch offices abroad, but Singer declined, and so Remington was forced into opening branch offices in most of the major European cities in the early years of the twentieth century.

By 1890 the Remington Typewriter Company had become a very large concern indeed, making 20,000 machines a year. During the last decade of the century it was joined by several competitors: Smith Premier in 1889, Oliver in 1894, Underwood in 1895, and Royal a few years later. During this period the typewriter as we know it became routinely available to large and small businesses. By 1900 there were at least a dozen major manufacturers making 100,000 typewriters a year. Sales continued to soar right up to World War I, by which time there were several dozen American firms in existence and more in Europe. Selling for an average price of $75, typewriters became the most widely used business machines, accounting for half of all office appliance sales.

The manufacture and distribution of typewriters made up, however, only half the story. Just as important was the training of workers to use them. Without training, a typewriter operator was not much more effective than an experienced writing clerk. Beginning in the 1880s, organizations came into being to train young workers (mostly women) for the new occupation of "type-writer," later known as a "typist." Typing had a good deal in common with stenography and telegraphy, in that it took several months of intense practice to acquire the skill. Hence many typewriting schools grew out of the private stenography and telegraph schools that had existed since the 1860s. The typewriter schools helped satisfy a massive growth in the demand for trained office workers in the 1890s. Soon the public schools joined in the task of training this new workforce, equipping hundreds of thousands of young women and men with typewriting skills. The shortage of male clerks to fill the burgeoning offices at the turn of the century created the opportunity for women to enter the workplace in ever increasing numbers. By 1900 the U.S. Census recorded 112,000 typists and stenographers in the nation, of whom 86,000 were female. Twenty years previously, there had been just 5,000, only 2,000 of them women.

By the 1920s office work was seen as predominantly female, and typing as universally female. As one feminist author characterized the situation with intentional irony, "a woman's place is at the typewriter." Certainly the typewriter was the machine that above all brought women into the office, but there was nothing inherently gendered about the technology. As we write, a new generation of office workers—both male and female—is coming to terms with this fact.

For many years historians of information technology neglected the typewriter as a progenitor of the computer industry. But now we can see that it is important to the history of computing in that it pioneered three key features of the office-machine industry and the computer industry that succeeded it: the perfection of the product and low-cost manufacture; a sales organization to sell the product; and a training organization to enable workers to use the technology.

The Rands

Left to itself, it is unlikely that the Remington Typewriter Company would have succeeded in the computer age—certainly none of the other typewriter companies did. However, in 1927 Remington became part of a conglomerate,

Remington Rand, organized by James Rand and his son James Jr. The Rands were the inventor-entrepreneur proprietors of the Rand Kardex Company, the world's leading supplier of record-keeping systems.

Filing systems for record keeping were one of the breakthrough business technologies, occurring roughly in parallel with the development of the typewriter. The fact that loose-leaf filing systems are so obviously low-tech does not diminish their importance. Without the technology to organize loose sheets of paper, the typewriter revolution itself could never have occurred.

Before the advent of the typewriter and carbon copies, business correspondence was kept in a "letter book." There was usually both an ingoing and an outgoing letter book, and it was the duty of a copying clerk to make a permanent record of each letter as it was sent out or received—by copying it out in longhand into the letter book. It was inevitable that this cumbersome business practice would not survive the transition from small businesses to major enterprises in the decades after the Civil War. That process was greatly accelerated by the introduction of the typewriter and carbon copies, which made document preparation convenient.

The first modern "vertical" filing system was introduced in the early 1890s, and won a Gold Medal at the 1893 World's Fair. The vertical filing system was a major innovation, but it has become so commonplace that its significance is rarely noted. It is simply the system of filing records in wood or steel cabinets, each containing three or four deep drawers; in the drawers papers are stored on their edges. Papers on a particular subject or for a particular correspondent are grouped together between cardboard dividers with index tabs. Vertical filing systems used only a tenth of the space of the box-and-drawer filing systems they replaced, and made it much faster to access records.

At first vertical filing systems coped very well, but as business volumes began to grow in the 1890s, they became increasingly ineffective. One problem was that organizing correspondence or an inventory worked quite well for a hundred or even a thousand records, but became very cumbersome with tens or hundreds of thousands of records.

Around this time, James Rand Sr. (1859–1944) was a young bank clerk in his hometown of Tonawanda, New York. Fired up by the difficulties of retrieving documents in existing filing systems, in his spare time he invented and patented a "visible" index system of dividers, colored signal strips, and tabs, enabling documents to be held in vertical files and located through an index system three or four times faster than previously. The system could be expanded indefinitely, and it enabled literally millions of documents to be organized with absolute accuracy and retrieved in seconds.

In 1898 Rand organized the Rand Ledger Company, which quickly came to dominate the market for record-keeping systems in large firms. By 1908 the firm had branch offices in nearly forty U.S. cities and agencies throughout the world. Rand took his son, James Rand Jr., into the business in 1908. Following his father's example, Rand junior invented an ingenious card-based record-keeping system that the company began to market as the Kardex System. It combined the advantages of an ordinary card index with the potential for un-limited expansion and instant retrieval. In 1915 Rand junior broke away from his father and, in a friendly family rivalry, set up as the Kardex Company. Rand junior was even more successful than his father. By 1920 he had over a hundred thousand customers, ninety branch offices in the United States, a plant in Germany, and sixty overseas offices.

In 1925 Rand senior was in his mid-sixties and approaching retirement. At the urging of his mother, Rand junior decided that he and his father should resume their business association, and the two companies were merged as the Rand Kardex Company, with the father as chairman and the son as president and general manager. The combined firm was by far the largest supplier of business record-keeping systems in the world, with 4,500 field representatives in 219 branch offices in the United States and 115 agencies in foreign countries.

Rand junior set out on a vigorous series of mergers and business takeovers, aimed at making the company the world's largest supplier of all kinds of office equipment. Several other manufacturers of loose-leaf card index systems were acquired early on, followed by the Dalton Adding Machine Company, and the Powers Accounting Machine Company—the makers of a rival punched-card system to Hollerith's. In 1927, the company merged with the Remington Typewriter Company and became known as Remington Rand. With assets of $73 million, it was far and away the largest business-machine company in the world.

The First Adding Machines

Typewriters aided the documentation of information, while filing systems facilitated its storage. Adding machines were concerned with *processing* information. The development of mass-produced adding machines followed the development of the typewriter by roughly a decade, and there are many parallels in the development of the two industries.

The first commercially produced adding machine was the Arithmometer developed by Thomas de Colmar of Alsace at the early date of 1820. However, the Arithmometer was never produced in large quantities, and was hand-built probably at the rate of no more than one or two a month. The Arithmometer was not a very reliable machine, and it was a generation before it was produced in significant quantities (there is virtually no mention of it until the Great Exhibition of 1851 in London). Sturdy, reliable Arithmometers became available in the 1880s, but the annual production never exceeded a few dozen.

Although the Arithmometer was relatively inexpensive, costing about $150, demand remained low because its speed of operation was too slow for routine adding in offices. It had to be operated by "dialing" the number onto a set of wheels using a stylus and then actuated with a hand crank. Arithmometers were useful in insurance companies and in engineering concerns, where calculations of seven-figure accuracy were necessary. But they were irrelevant to the needs of ordinary bookkeepers. The bookkeeper of the 1880s was trained to add up a column of four-figure amounts mentally almost without error, and could do so far more quickly than using an Arithmometer.

At the beginning of the 1880s the critical challenge in designing an adding machine for office use was to speed up the rate at which numbers could be entered. Of course, once this problem had been solved a second problem came into view: Financial organizations, particularly banks, needed a written record of numbers as they were entered into the adding machine so that they would have a permanent record of their financial transactions.

The solvers of these two critical problems, Dorr E. Felt and William S. Burroughs, were classic inventor-entrepreneurs. The machines they created, the Comptometer and the Burroughs Adding Machine, dominated the worldwide market well into the 1950s. Only the company founded by Burroughs, however, successfully made the transition into the computer age.

Dorr E. Felt was an obscure machinist living in Chicago when he began to work on his adding machine. The technological breakthrough that he achieved in his machine was that it was "key-driven." Instead of the dials of the Arithmometer, or the levers used on other adding machines that had to be painstakingly set, Felt's adding machines had a set of keys, a little like those of a typewriter. The keys were arranged in columns, labeled 1 through 9 in each column. If an operator pressed one figure in each column, say, the 7, 9, 2, 6, and 9 keys, the amount $792.69 would be added into the total displayed by the machine. A dexterous operator could enter a number with as many as ten digits (that is, one digit per finger) in a single operation. In effect, Felt had done for the adding machine what Sholes had done for the typewriter.

In 1887, at the age of twenty-four, Felt went into partnership with a local manufacturer, Robert Tarrant, to manufacture the machine as the Felt & Tarrant Manufacturing Company. As with the typewriter, sales were slow at first, but by 1900 the firm was selling over a thousand machines a year.

There was, however, more to the Comptometer business than making and selling machines. As with the typewriter, the skill to use the Comptometer was not easily acquired. Hence the company established Comptometer training schools. By World War I schools had been established in most major cities of the United States, as well as in several European centers. At the Comptometer schools young women and men, fresh from high school, would in a period of a few months of intense training become adept at using the machines. Operating a Comptometer at speed was an impressive sight: In film archives one can find silent motion pictures showing Comptometer operators entering data far faster than it could be written down, far faster than even the movies could capture—their fingers appear as a blur. Like learning to ride a bicycle, once the skill was acquired it would last a working life—until Comptometers became obsolete in the 1950s.

Between the two world wars, the Comptometer was phenomenally success-ful, and millions of them were produced. But the company never made it into the computer age. This was essentially because Felt & Tarrant failed to develop advanced electromechanical products in the 1930s. When the vacuum tube computer arrived after World War II, the company was unprepared to make the technological leap into computers. Eventually it was acquired in the 1960s in a complex merger and, like much of the old office-appliance industry, vanished into oblivion.

William S. Burroughs made the first successful attack on the second critical problem of the office adding machine: the need to print its results. Burroughs was the son of a St. Louis mechanic and was employed as a bank clerk, where the drudgery of long hours spent adding up columns of figures is said to have caused a breakdown in his health. He therefore decided at the age of twenty-four to quit banking and follow his father by becoming a mechanic. Putting the two together, Burroughs began to develop an adding machine for banks, which would not only add numbers rapidly like a Comptometer, but would also print the numbers as they were entered.

In 1885 Burroughs filed his first patent, founded the American Arithmome-ter Company, and began to hand-build his "adder-lister." The first businesses targeted by Burroughs were the banks and clearing houses for which the ma-chine had been especially designed. After a decade's mechanical perfection and gradual buildup of production, by 1895 the firm was selling several hundred machines a year, at an average price of $220. When William Burroughs died in his mid-forties in 1898, the firm was still a relatively small organization, selling fewer than a thousand machines in that year.

In the early years of the new century, however, sales began to grow very rapidly. By 1904 the company was making 4,500 machines a year, had outgrown its St. Louis plant, and had moved to Detroit, where it was renamed the Burroughs Adding Machine Company. Within a year annual production had risen to nearly 8,000 machines, the company had 148 representatives selling the machines, and several training schools had been established. Three years later Burroughs was selling over 13,000 machines a year, and its advertisements boasted that there were 58 different models of machine: "One Built for Every Line of Business."

In the first decade of the new century, Felt & Tarrant and the Burroughs Adding Machine Company were joined by dozens of other calculating-machine manufacturers such as Dalton (1902), Wales (1903), National (1904), and Madas (1908). A major impetus to development of the U.S. adding-machine industry occurred in 1913 with the introduction of a new tax law that adopted progressive tax rates and the withholding of tax from pay. A further increase in paperwork occurred during World War I, when corporate taxes were expanded.

Of all the many American adding-machine firms that came into existence before World War I, only Burroughs made a successful transition into the computer age. One reason was that Burroughs supplied not only adding machines but also the intangible know-how of incorporating them into an existing business organization. The company became very adept at this, and effectively sold business systems alongside their machines. Another factor was that Burroughs did not cling to a single product, but gradually expanded its range in response to user demand. Between the two world wars Burroughs moved beyond adding machines and introduced full-scale accounting machines. The knowledge of business accounting systems became deeply embedded in the sales culture of Burroughs, and this helped ease its transition to computers in the 1950s and 1960s.

The National Cash Register Company

In this brief history of the office-machine industry, we have repeatedly alluded to the importance to firms of their sales operations, in which hard selling was reinforced by the systematic analysis of customer requirements, the provision of after-sales support, and user training. More than any other business sector, the office-machine industry relied on these innovative forms of selling, which were pioneered largely by the National Cash Register Company in the 1890s. When an office-machine firm needed to build up its sales operation, the easiest way was to hire someone who had learned the trade with NCR. In this way the techniques that NCR developed diffused throughout the office-machine industry and, after World War II, became endemic to the computer industry.

Several histories have been written of NCR, and they all portray John H. Patterson, the company's founder, as an aggressive, egotistical crank and identify him as the role model for Thomas J. Watson Sr., the future leader of IBM. At one point in his life, Patterson took up Fletcherism, a bizarre eating fad in which each mouthful of food was chewed sixty times before swallowing. Another time he became obsessed with calisthenics and physical fitness and took to horse riding at six o'clock each morning—and obliged his senior executives to accompany him. He was capricious, as likely to reward a worker with whom he was pleased with a large sum of money as to sack another on the spot for a minor misdemeanor. He was a leading figure in the Garden City movement, and the NCR factory in Dayton, Ohio, was one of its showpieces, its stream-lined buildings nestling in a glade planted with evergreen trees. However, for all his idiosyncrasies, there is no question that Patterson possessed an instinct for organizing the cash register business that was so successful that NCR's "business practices and marketing methods became standard at IBM and in other office appliance firms by the 1920s."

As with the typewriter and the adding machine, there were a number of attempts to develop a cash register during the nineteenth century. The first practical cash register was invented by a Dayton restaurateur named James Ritty. Ritty believed he was being defrauded by his staff, so he invented "Ritty's Incorruptible Cashier" in 1879. His machine operated very much like a modern cash register: When a sale was made, the amount recorded was displayed prominently for all to see, as a check on the clerk's honesty. Inside the machine, the transaction was recorded on a roll of paper. At the end of the day, the store owner would add up the day's transactions and check the total against the cash received. Ritty tried to go into business manufacturing and selling his machine, but was unsuccessful and sold out to a local entrepreneur. Apparently, before giving up Ritty sold exactly one machine—to John H. Patterson, who was then a partner in a coal retailing business.

In 1884, at the age of forty, Patterson, having been outmaneuvered by some local mining interests, decided to quit the coal business, even though "the only thing he knew anything about was coal." This was not quite true: He also knew about cash registers, since he was one of the very few users of the machine. He decided to use some of the money from the sale of the coal business to buy the business originally formed by Ritty, which he renamed the National Cash Register Company. Two years later the National Cash Register Company was selling more than a thousand machines a year.

Patterson understood that the cash register needed constant improvement to keep it technically ahead of the competition. In 1888 he established a small "inventions department" for this purpose. This was probably the first formal research and development organization to be established in the office-machine industry, and it was copied by IBM and others. Over the next forty years, the NCR inventions department was to accrue more than two thousand patents. Its head, Charles Kettering, went on to found the first research laboratory for General Motors.

With Patterson's leadership, NCR came to dominate the world market for cash registers and grew phenomenally. By 1900 NCR was selling nearly 25,000 cash registers a year and had 2,500 employees. By 1910 it was selling 100,000 registers a year and had more than 5,000 employees. In 1922, the year Patterson died, NCR sold its two-millionth cash register.

Under Patterson, NCR had been a one-product company, but however successful that product had been, NCR would never have become a major player in the computer industry. But after Patterson's death, the executives who remained in control of NCR decided to diversify into accounting machines. As Stanley Allyn, later a president of the company, put it, NCR decided to "turn off the worn, unexciting highway into the new land of mechanized accounting."

By the early 1920s, NCR had very sophisticated market research and technical development departments. Between them they initiated a major redirection of the company's business, based on the Class 2000 accounting machine launched in 1926. A fully fledged accounting machine, the Class 2000 could handle invoicing, payroll, and many other business accounting functions. From this time on, the company was no longer called the National Cash Register Company but was known simply as NCR. The NCR Class 2000 accounting machine was as sophisticated as any accounting machine of its day and was fully equal to the machines produced by Burroughs.

NCR's greatest legacy to the computer age, however, was the way it shaped the marketing of business machines and established virtually all of the key sales practices of the industry. It would be wrong to credit Patterson with inventing all of these sales techniques. Certainly he invented some basic ideas, but mostly he embellished and formalized existing practices. For example, he adopted the Singer Sewing Machine idea of establishing a series of branch retail outlets.

Cash registers could be bought directly from the NCR outlets, and if they broke down they could be returned for repair by factory-trained engineers, with a loaner machine provided while the machine was being mended.

Patterson learned early on that store owners rarely walked off the street into an NCR store to buy a cash register; as late as the turn of the century, few retailers even knew what a cash register was. As Patterson put it, cash registers were sold not bought. Thus he developed the most effective direct sales force in the country.

Patterson knew from personal experience that selling could be a lonely and soul-destroying job. Salesmen needed to be motivated both financially and spiritually. Financially, Patterson rewarded his salesmen with an adequate basic salary and a commission that could make them positively rich. In 1900 he introduced the concept of the sales quota, which was a stick-and-carrot sales incentive. The carrot was membership in the Hundred Point Club, for those achieving 100 percent of their quota. The reward for membership in the club was the prestige of attending an annual jamboree at the head office in Dayton, being recognized by the company brass, and hearing an inspirational address from Patterson himself. He would illustrate his talks using an oversized notepad resting on an easel, on which he would crayon the main points of his argument. As NCR grew in size, the Hundred Point Club grew to several hundred members, far exceeding the capacity of Dayton's hotels to accommodate them, so the annual get-together was moved to a tent city on a grassy field owned by NCR.

NCR salesmen were supported by a flood of literature produced by the NCR head office. From the late 1880s, NCR began to mail product literature to tens of thousands of sale "prospects" (an NCR term). At first these pamphlets were single broadsheets, but later a magazine-style booklet called *The Hustler* was produced, which targeted specific merchants such as dry-goods stores or saloons. These booklets showed how to organize retail accounting systems and gave inspirational encouragement: "You insure your life. Why not insure your money too! A National cash register will do it."

Patterson also established sales training. He transformed a career in selling from being a somewhat disreputable occupation to something close to a profession. In 1887 he got his top salesman to set down his sales patter in a small booklet, which he called the *NCR Primer*. Effectively a sales script, it was "the first canned sales approach in America." In 1894 Patterson founded a sales training school through which all salesmen eventually passed: "The school taught the primer. It taught the demonstration of all the machines the company then made. It taught price lists, store systems, the manual, the getting of prospects, and the mechanism of the machine." An early graduate of the NCR sales school was Thomas J. Watson. Attending the school changed his life.

Thomas Watson and the Founding of IBM

Thomas J. Watson was born in Campbell, New York, in 1874, the son of a strict Methodist farmer. At the age of eighteen, Watson went to commercial college, then became a bookkeeper, earning six dollars a week. He disliked sedentary office life, however, and decided to take a riskier but better-paid (ten-dollar-a-week) job selling pianos and organs. On the road, with a horse-drawn wagon, carrying a piano, he learned to sell. These were basic if naive lessons in developing a sales patter and a glad-handing approach.

Always looking to better himself, in 1895 Watson managed to get a post as a cash register salesman with NCR. He was then one among hundreds of NCR salesmen. Armed with the *NCR Primer*—and assiduously learning from established salesmen—Watson became an outstanding salesman. He attended the sales training school at NCR's headquarters and claimed that it enabled him to double his sales thereafter. After four years as one of NCR's top salesmen, he was promoted to overall sales agent for Rochester, New York. At the age of twenty-nine, he had a record that marked him out for higher things.

In 1903 Patterson summoned Watson to his Dayton headquarters and offered him the opportunity to run an "entirely secret" operation. At this time NCR was facing competition from sellers of second-hand cash registers. The proposal that Patterson put to Watson was that he should buy up second-hand cash registers, sell them cheaply, and hence drive the second-hand dealers out of business. As Patterson put it, "the best way to kill a dog is to cut off its head." Even by the buccaneering standards of the early twentieth century this was an unethical, not to say illegal, operation. Watson later came to regret this moral lapse, but it served to push him upwards in NCR.

At age thirty-three, Watson in 1908 was promoted to sales manager, and in that position he gained an intimate knowledge of what was acknowledged to be the finest sales operation in America. He found himself in charge of the company's sales school and, like Patterson, he had to give inspirational talks. Watson exceeded even Patterson as the master of the slogan: "ever onward . . . aim high and think in big figures; serve and sell; he who stops being better stops being good." And so on. Watson coined an even simpler slogan, the single word T-H-I-N-K. Soon notices bearing the word were pinned up in NCR branch offices.

By 1911 Watson had become heir apparent as general manager of NCR—when the capricious Patterson fired him. The thirty-nine-year-old supersalesman Thomas Watson found himself unemployed. He would shortly surface as the president of C-T-R, the company that had acquired the rights to the Hollerith punched-card system.

▪ ▪ ▪

By about 1905, although the typewriter and adding-machine industries had become significant sectors of the economy, the punched-card machine industry was still in its infancy, for there were no more than a handful of installations in the world, compared with over a million typewriters and hundreds of thousands of adding machines. Twenty years later this picture would be dramatically altered.

When Hollerith first went into business with his electric tabulating system in 1886, it was not really a part of the office-machine industry in the same sense as Remington Typewriter, Burroughs Adding Machine, or NCR. These firms were characterized by the high-volume manufacture of relatively low-priced machines. By contrast, Hollerith's electric tabulating system was a very expensive, highly specialized system limited in its applications to organizations such as the Census Bureau. Hollerith lacked the vision and background to turn his company into a mainstream office-machine supplier, and it would never have become a major force in the industry until he passed control to a manager with a business-machine background.

The 1890 U.S. population census was a major triumph for Hollerith. But as the census started to wind down in 1893, he was left with a warehouse full of census machines—and there most of them would remain until the 1900 census. In order to maintain the income of his business he needed to find new customers, so he tried to adapt his machines so that they could be used by ordinary American businesses.

On 3 December 1896, Hollerith incorporated his business as the Tabulating Machine Company (TMC). He managed to place one set of machines with a railway company, but he was quickly diverted by the much easier profits of the upcoming 1900 U.S. population census. The census sustained the Tabulating Machine Company for another three years, but as the census operation wound down, Hollerith was again faced with a declining income and turned his attention once more to private businesses. This time he did so with much more application, however, and managed to place several sets of machines with commercial organizations.

It was just as well that Hollerith had turned his attention to the commercial applications of his machines, at last, because his easy dealings with the Census Bureau were about to come to an end. The superintendent of the census was a political appointment, and Superintendent Robert Porter—who had supported Hollerith in both the 1890 and 1900 censuses—was the appointee of President McKinley (whose biography Porter wrote). In 1901, McKinley was assassinated and Porter's career in the public service came to an abrupt end. Porter returned to his home country of England and set up the British Tabulating Machine Company in London. Meanwhile a new superintendent had been appointed, one who did not care for the Census Bureau's cozy relationship with Hollerith's company and who considered the charges for the 1900 census to have been unacceptably high.

In 1905, unable to agree on terms, Hollerith broke off business relations with the Census Bureau and put all his efforts into the commercial development of his machines. Two years later the Bureau engaged a mechanical engineer named James Powers to improve and develop tabulating machines for the next census. Powers was an inventor-entrepreneur in the same mold as Hollerith, and he was soon to emerge as a serious competitor—producing a machine that improved on Hollerith's by printing its results.

Meanwhile Hollerith continued to perfect the commercial machines, producing a range of "automatics" that was to stay in production for the next twenty years. There were three punched-card machines in the commercial setup: a keypunch to perforate the cards, a tabulator to add up the numbers punched on the cards, and a sorter to arrange the cards in sequence. All punched-card offices had at least one each of these three machines. Larger offices might have several of each, especially the keypunches. It was common to have a dozen or more female keypunch operators in a large office. The automatics made use of a new "45-column" card, $7\frac{1}{2} \times 3\frac{1}{4}$ inches in size, on which it was possible to store up to 45 digits of numerical information. This card became the industry standard for the next twenty years.

With the introduction of the automatic machines, TMC raced ahead. By 1908 Hollerith had thirty customers. During the next few years the company grew at a rate of over 20 percent every half-year. By 1911 it had approximately a hundred customers and had completely made the transition to being an office-machine company. A contemporary American journalist captured exactly the extent of the booming punched-card business:

> The system is used in factories of all sorts, in steel mills, by insurance companies, by electric light and traction and telephone companies, by wholesale

merchandise establishments and department stores, by textile mills, automobile companies, numerous railroads, municipalities and state governments. It is used for compiling labor costs, efficiency records, distribution of sales, internal requisitions for supplies and materials, production statistics, day and piece work. It is used for analyzing risks in life, fire and casualty insurance, for plant expenditures and sales of service, by public service corporations, for distributing sales and cost figures as to salesmen, department, customer, location, commodity, method of sale, and in numerous other ways. The cards, besides furnishing the basis for regular current reports, provide also for all special reports and make it possible to obtain them in a mere fraction of the time otherwise required.

All of these punched-card installations lay at the heart of sophisticated information systems. Hollerith abhorred what he perceived as the sharp selling practices of the typewriter and adding-machine companies, and he personally—or, increasingly, one of his assistants—worked closely with customers to integrate the punched-card machines into the overall information system.

By 1911 Herman Hollerith was the well-to-do owner of a successful business, but he was fifty-one and his health was failing. He had always refused to sell the business in the past, or even to share the running of it; but his doctor's advice and generous financial terms finally overcame his resistance.

An offer to take over TMC came from one of the leading business promoters of the day, Charles Ranlegh Flint, the so-called "father of Trusts." Flint was probably the most celebrated exponent of the business merger. He proposed merging the Tabulating Machine Company with two other firms—the Computing Scale Company, a manufacturer of shopkeeper's scales, and the International Time Recording Company, which produced automatic recording equipment for clocking employees in and out of the workplace. Each of these companies was to contribute one word to the name of a new holding company, the Computing-Tabulating-Recording Company, or C-T-R.

In July 1911, the Tabulating Machine Company was sold for a sum of $2.3 million, of which Hollerith personally received $1.2 million. Now a wealthy man, Hollerith went into semiretirement, although, never quite able to let go, he remained a technical consultant for another decade but made no further important contributions to punched-card machinery. Hollerith lived to see C-T-R become IBM in 1924, but he died on the eve of the stock market crash of 1929—and so never lived to see the Great Depression, nor how IBM survived it and became a business-machine legend.

There were three reasons for the ultimate success of IBM: the development of a sales organization based on that of NCR; the "rent-and-refill" nature of the punched-card machine business; and technical innovation. All of these were intimately bound up with the appointment of Thomas J. Watson Sr. as C-T-R's general manager in 1911 and its president in 1914.

After being fired from NCR by Patterson in 1911, Watson was not short of offers. However, he was keen to be more than a paid manager and wanted to secure a lucrative, profit-sharing deal. He therefore decided to accept an offer from Charles Flint to become general manager of C-T-R. While C-T-R was a minute organization compared with NCR, Watson understood—far more than Flint or Hollerith—the potential of the Hollerith patents. He therefore negotiated a small base salary plus a commission of 5 percent of the profits of the company. If C-T-R grew in the way that Watson intended, it would eventually make him the highest-paid businessman in America—and by 1934, with a salary of $364,432, that was what he had officially become.

Making a break with Hollerith's low-key selling methods, Watson immediately introduced wholesale the sales practices he had seen so successfully pioneered in NCR. He established sales territories, commissions, and quotas. Where NCR had its Hundred Point Club for salesmen who achieved their sales quotas, Watson introduced the One Hundred Per Cent Club in C-T-R. In a period of about five years, Watson completely transformed the culture and prospects of C-T-R, turning it into the brightest star of the office-machine industry. By 1920 C-T-R's income had more than tripled, to $15.9 million. Subsidiaries were opened in Canada, South America, Europe, and the Far East. In 1924 Watson renamed the company International Business Machines and announced, "Everywhere . . . there will be IBM machines in use. The sun never sets on IBM."

Watson never forgot the lessons he had learned from Patterson. As a writer in *Fortune* magazine observed in 1932:

> [L]ike a gifted plagiarist, a born imitator of effective practices of others, [Watson] fathomed the system and excelled over all. . . . It was Patterson's invariable habit to wear a white stiff collar and vest; the habit is rigorously observed at I.B.M. It was Patterson's custom to mount on easels gigantic scratch-pads the size of newspapers and on them to record, with a fine dramatic flourish, his nuggets of wisdom; nearly every executive office at I.B.M. has one. It was Patterson's custom to expect all employees to think, act, dress, and even eat the way he did; all ambitious I.B.M. folk respect many of Mr. Watson's

predilections. It was Patterson's custom to pay big commissions to big men, to meet every depression head on by increasing his selling pressure; these are part of the I.B.M. technique.

The T-H-I-N-K motto was soon to be seen hanging on the wall in every office in the IBM empire.

If IBM was famous for its dynamic salesmen, it became even more celebrated for its apparent immunity to business depressions. The "rent-and-refill" nature of the punched-card business made IBM virtually recession-proof. Because the punched-card machines were rented and not sold, even if IBM acquired no new customers in a bad year, its existing customers would continue to rent the machines they already had, ensuring a steady income year after year. The rental of an IBM machine repaid the manufacturing costs in about two to three years, so that after that period virtually all the income from a machine was profit. Most machines remained in service for at least seven years, often ten, and occasionally fifteen or twenty years. Watson fully understood the importance of IBM's leasing policy to its long-term development and—right into the 1950s—resisted pressure from government and business to sell IBM machines outright.

The second source of IBM's financial stability was its punched-card sales. The cards cost only a fraction of the dollar per thousand for which they were sold. A 1930s journalist explained that IBM belonged to a special class of business:

> a type that might be called "refill" businesses. The use of its machines is followed more or less automatically and continually by the sale of cards. This type of business also has good precedents: in the Eastman Kodak Co., which sells films to camera owners; in General Motors, whose AC spark plugs are sold to motor-car owners; in Radio Corp., whose tubes are bought by radio owners; in Gillette Razor Co., whose blades are bought by razor users.

In the case of razor blades and spark plugs (somewhat less in the case of photographic film and radio tubes), manufacturers must compete with one another for the refill market. IBM, however, found itself secure in the market of tabulator cards for its machines, because the cards had to be made with great accuracy on a special paper stock which was never successfully imitated.

By the 1930s, IBM was selling 3 billion cards a year, which accounted for 10 percent of its income but 30 to 40 percent of its profits.

Technical innovation was the third factor that kept IBM at the forefront of the office-machine industry between the two world wars. In part, this was a response to competition from its major rival—the Powers Accounting Machine Company, which had been formed in 1911 by James Powers, Hollerith's successor at the Census Bureau. Almost as soon as he arrived at C-T-R, Watson saw the need for a response to the Powers printing machine and set up an experimental department similar to the inventions department Patterson had set up at NCR, to systematically improve the Hollerith machines. The printing tabulator development was interrupted by World War I, but in the first postwar sales conference in 1919:

> Watson had one of the supreme moments of his career. The curtains behind the central podium were drawn back, Watson threw a switch, and the new machine standing at the centre of the stage began printing results as cards flowed through it. Salesmen stood up in the chairs and cheered, exuberant that they were going to be able to match the competition at last.

In 1927 Powers was acquired by Remington Rand, which had strong organizational capabilities, and the competitive threat to IBM increased significantly. The following year IBM brought out an 80-column card that had nearly twice the capacity of the 45-column card that both companies were then selling. Remington Rand countered two years later with a 90-column card. And so it went. This pattern of development, each manufacturer leapfrogging the other, characterized the industry both in the United States and in the rest of the developed world—where IBM and Remington Rand both had subsidiaries, and where there were other European punched-card machine manufacturers.

In the early 1930s IBM leapfrogged the competition by announcing the 400 Series accounting machines, which marked a high point in interwar punched-card machine development. The most important and profitable machine of the range was the model 405 Electric Accounting Machine. IBM was soon turning out 1,500 copies of this machine a year, and it was "the most lucrative of all IBM's mechanical glories." The 400 Series was to remain in manufacture until the late 1960s, when punched-card machines finally gave way completely to computers. Containing 55,000 parts (2,400 different) and 75 miles of wire, the model 405 was the result of half a century of evolution of the punched-card machine art. The machine was incomparably more complex than an ordinary accounting machine, such as the NCR Class 2000, which contained a "mere" 2,000 parts. Thus by the 1930s IBM had a significant technical lead over its competitors, which eased its transition to computers in the 1950s.

Powers always came a poor second to IBM. This remained true even after it was acquired in the Remington Rand merger in 1927. While IBM's machines

on, daughter of the poet. A young gentlewoman aged about eighteen,
eloped an interest in mathematics when such a calling was very un-
oman. She managed to charm some of England's leading mathemati-
ding Babbage and Augustus de Morgan—into guiding her studies.
aid Babbage by writing the best nineteenth-century account of his
ngine.

never built a full-scale Difference Engine because in 1833 he aban-
a new invention, the Analytical Engine, the chief work on which
the history of computing rests. He was at that time at the very
powers: Lucasian Professor of Mathematics at Cambridge Univer-
f the Royal Society, one of the leading lights in scientific London,
of the most influential economics book of the 1830s, the *Economy
ures.*

tant as the Difference Engine had been, it was fundamentally a
eption in that all it could do was produce mathematical tables. By
Analytical Engine was to be a universal machine capable of any
al computation. The idea of the Analytical Engine came to Bab-
he was considering how to eliminate human intervention in the
ngine by feeding back the results of a computation, which he re-
the engine "eating its own tail." Babbage took this simple refine-
Difference Engine, and from it evolved the design of the Analytical
ch embodies almost all the important functions of the modern
uter.

significant concept in the Analytical Engine was the separation of
omputation from the storage of numbers. In the original Difference
two functions had been intimately connected: In effect, numbers
in the adding mechanisms, as in an ordinary calculating machine.
abbage had been concerned by the slow speed of his adding mecha-
he attempted to speed up by the invention of the "anticipatory car-
ethod of rapid adding that has its electronic equivalent in the
puter in the so-called carry-look-ahead adder. The extreme com-
e new mechanism was such that it would have been far too expen-
uilt any more adders than necessary, so he separated the arithmetic
storage functions. Babbage named these two functional parts of the
ill and the *store*, respectively. This terminology was a metaphor
the textile industry. Yarns were brought from the store to the mill,
vere woven into fabric, which was then sent back to the store. In the
ngine, numbers would be brought from the store to the arithmetic

were said to be more reliable and better maintained than those of Powers, IBM's salesmanship was overwhelmingly better. IBM's salesmen—or, rather, systems investigators—were ambassadors who had been trained to perfection in a special school in Endicott, New York. It was estimated in 1935 that IBM had more than 4,000 accounting machines in customer installations. This gave it a market share of 85 percent, compared with Remington Rand's 15 percent.

Watson became famous in the annals of American business history for his nerve during the Great Depression. During this period, business-equipment manufacturers generally suffered a 50 percent decline in sales. IBM suffered a similar decline in new orders, but its income was buttressed by card sales and the rental income from machines in the field. Against the advice of his board, Watson held on to his expensively trained workforce and maintained IBM's factories in full production, building up frighteningly high stocks of machines in readiness for the upturn. Watson's confidence was a beacon of hope in the Depression years, and it turned him into one of the most influential businessmen in America. He became president of the American Chamber of Commerce, and later of the International Chamber of Commerce, and he was an adviser and friend of President Franklin D. Roosevelt's.

The recovery for IBM, and the nation, came with Roosevelt's New Deal in 1935, of which Watson was a prominent supporter. Under the Social Security Act of 1935, it became necessary for the federal government to maintain the employment records of the entire working population of 26 million people. IBM, with its overloaded inventories and with factories in full production, was superbly placed to benefit.

In October 1936, the federal government took delivery of a large slice of IBM's inventory—415 punches and accounting machines—and located the equipment in a brick-built Baltimore loft building of 120,000 square feet. In what was effectively an information production line, half a million cards were punched, sorted, and listed every day. Nothing like it had been seen since the days of the 1890 census. IBM's income for "the world's biggest bookkeeping job" amounted to $438,000 a year—which was only about 2 percent of IBM's income, but soon government orders would count for 10 percent. Moreover, New Deal legislation had created obligations for employers in the private sector to comply with federal demands for information on which welfare, National Recovery Act codes, and public works projects were dependent. This fueled demand further. Between 1936 and 1940, IBM's sales nearly doubled, from $26.2 million to $46.2 million, and its workforce rose to 12,656. By 1940 IBM was doing more business than any other office-machine firm in the world. With sales of less than $50 million, it was still a relatively small firm, but the future seemed limitless. No wonder that a journalist could write in *Fortune* magazine:

> Most enterprises, when they would dream of the exaltations of a golden age, roll their eyes backward; the International Business Machines Corp. has beheld no past so golden as the present. The face of Providence is shining upon it, and clouds are parted to make way for it. Marching onward as to war, it has skirted the slough of depression and averted the quicksands of false booms. Save for a few lulls that may be described as breathing spells, its growth has been strong and steady.

And the best was yet to come.

3

■■ BABBAGE'S DREAM COMES TRUE

IN OCTOBER 1946 the computing pionee
British readers in the science news journal *Na*
to support the construction of Babbage's calcu

> The black mark earned by the government o
> years ago for its failure to see Charles Babbag
> a successful conclusion has still to be wiped o
> it cost Britain the leading place in the art of m

> The reason for Comrie's remark was the c
> matic computing machine at Harvard Universi
> Dream Come True." Comrie, a New Zealander
> sity of Auckland in 1916, was one of many pe
> had accepted at face value Babbage's claim that
> gines was the fault of the British government. E
> plicated than this.
> Babbage invented two calculating machines
> Analytical Engine. Of the two, the Difference E
> nically, the less interesting—although it was th
> actually being built. Babbage worked on his I
> mately a decade, starting in the early 1820s. Th
> the model Difference Engine of 1833, described
> this model in the drawing room of his London h
> a conversation piece for his fashionable soirées
> were incapable of understanding what the mach

was Ada B
Ada had d
usual in a
cians—in
She later
Analytica
Babba
doned it
his fame
height of
sity, fello
and auth
of Manuf
As im
limited c
contrast,
mathem
bage wh
Differen
ferred te
ment of
Engine,
digital c
The
arithme
Engine,
were st
Origina
nisms,
riage"–
moder
plexity
sive to
and nu
engine
drawn
where
Analyt

mill for processing, and the results of the computation would be returned to the store. Thus, even as he wrestled with the complexities of the Analytical Engine, Babbage the economist was never far below the surface.

For about two years Babbage struggled with the problem of organizing the calculation—the process we now call programming, but for which Babbage had no word. After toying with various mechanisms, such as the pegged cylinders of barrel organs, he hit upon the Jacquard loom. The Jacquard loom, invented in 1802, had started to come into use in the English weaving and ribbon-making industries in the 1820s. It was a general-purpose device, which, when instructed by specially punched cards, could weave infinite varieties of pattern. Babbage envisaged just such an arrangement for his Analytical Engine.

Meanwhile, the funding difficulties of the original Difference Engine were still not resolved, and Babbage was asked by the Duke of Wellington—then prime minister—to prepare a written statement. It was in this statement, dated 23 December 1834, that Babbage first alluded to the Analytical Engine—"a totally new engine possessing much more extensive powers." It may be that Babbage was aware that raising the subject of the Analytical Engine would weaken his case, but he claimed in his statement that he was merely doing the honorable thing by mentioning the new engine, so that "you may have fairly before you all the circumstances of the case." Nonetheless, the clear messages the government read into the statement were, first, that it should abandon the Difference Engine—and the £17,470 it had already spent on it—in favor of Babbage's new vision; and, second, that Babbage was more interested in building the engine than in making the tables that were the government's primary interest. As a result, Babbage fatally undermined the government's confidence in the project, and no decision on the engines was to be forthcoming for several years.

Between 1834 and 1846, without any outside, financial support, Babbage elaborated the design of the Analytical Engine. There was no question at this time of his attempting construction of the machine without government finance, so it was largely a paper exercise—eventually amounting to several thousand manuscript pages. During this period of intense creative activity, Babbage essentially lost sight of the objective of his engines and they became an end in themselves. Thus, while one can see the original table-making Difference Engine as completely at one with the economic need for nautical tables, with the Analytical Engine these issues were entirely forgotten. The Analytical Engine served no real purpose that the Difference Engine could not have satisfied—and it was an open question as to whether the Difference Engine itself was more effective than the more mundane team of human computers used by de Prony in the 1790s.

The fact is that, apart from Babbage himself, almost no one wanted or needed the Analytical Engine.

By 1840 Babbage was understandably in low spirits from the rebuffs he had received from the British government. He was therefore delighted when he was invited to give a seminar on his engines to the Italian Scientific Academy in Turin. The invitation reinforced his delusion of being a prophet without honor in his own country. The visit to Turin was a high point in Babbage's life, the crowning moment of which was an audience with King Victor Emmanuel, who took a very active interest in scientific matters. The contrast with his reception in England could not have been more marked.

In Turin, Babbage encouraged a young Italian mathematician, Lieutenant Luigi Menabrea, to write an account of the Analytical Engine, which was subsequently published in French in 1842. (Babbage chose well in Menabrea: He later rose to become prime minister of Italy.) Back in England, Babbage encouraged Ada—who was now in her late twenties and married to the Earl of Lovelace—to translate the Menabrea article into English. However, she went far beyond a mere translation of the article, adding lengthy notes of her own which almost quadrupled the length of the original. Her *Sketch of the Analytical Engine* was the only detailed account published in Babbage's lifetime—indeed, until the 1980s. Lovelace had a poetic turn of phrase, which Babbage never had, and this enabled her to evoke some of the mystery of the Analytical Engine that must have appeared truly remarkable to the Victorian mind:

> The distinctive characteristic of the Analytical Engine, and that which has rendered it possible to endow mechanism with such extensive faculties as bid fair to make this engine the executive right hand of abstract algebra, is the introduction into it of the principle which Jacquard devised for regulating, by means of punched cards, the most complicated patterns in the fabrication of brocaded stuffs. . . . We may say most aptly that the Analytical Engine *weaves algebraical patterns* just as the Jacquard loom weaves flowers and leaves.

This image of weaving mathematics was an appealing metaphor among the first computer programmers in the 1940s and 1950s.

One should note, however, that the extent of Lovelace's intellectual contribution to the *Sketch* has been much exaggerated. She has been pronounced the world's first programmer and even had a programming language (Ada) named in her honor. Recent scholarship has shown that most of the technical content and all of the programs in the *Sketch* were Babbage's work. But even if the

Sketch were based almost entirely on Babbage's ideas, there is no question that Lovelace provided its voice. Her role as the prime expositor of the Analytical Engine was of enormous importance to Babbage, and he described her, without any trace of condescension, as his "dear and much admired Interpreter."

Shortly after his return from Italy, Babbage once again began to negotiate with the British government for funding of the Analytical Engine. By this time there had been a change of government and a new prime minister, Robert Peel, was in office. Peel sought the best advice he could get, including that of the astronomer royal, who declared Babbage's machine to be "worthless." It was never put this baldly to Babbage, but the uncompromising message he received from the government was that he had failed to deliver on his earlier promise and that the government did not intend to throw good money after bad. Today, one can well understand the government's point of view, but one must always be careful in judging Babbage by twentieth-century standards. Today, a researcher would expect to demonstrate a measure of success and achievement before asking for further funding. Babbage's view, however, was that once he had hit upon the Analytical Engine, it was a pointless distraction to press ahead with the inferior Difference Engine merely to demonstrate concrete results.

By 1846 Babbage had gone as far as he was able with his work on the Analytical Engine. For the next two years he returned to the original Difference Engine project. He was now able to take advantage of a quarter of a century of improvements in engineering practice and many design simplifications he had developed for the Analytical Engine. He produced complete plans for what he called Difference Engine Number 2.

The government's complete lack of interest in the plans finally brought home to Babbage that all prospects of funding his engines were gone forever. The demise of his cherished project embittered him, although he had too much English reserve to express his disappointment directly. Instead, he turned to sweeping and sarcastic criticism of the English hostility to innovation:

> Propose to an Englishman any principle, or any instrument, however admirable, and you will observe that the whole effort of the English mind is directed to find a difficulty, a defect, or an impossibility in it. If you speak to him of a machine for peeling a potato, he will pronounce it impossible: if you peel a potato with it before his eyes, he will declare it useless, because it will not slice a pineapple.

Probably no other statement from Babbage so completely expresses his contempt for the bureaucratic mind.

As Babbage reached his mid-fifties, his intellectual powers were a shadow of what they had been twenty years earlier, and he was an increasingly spent and irrelevant force in the scientific world. He published the odd pamphlet and amused himself with semirecreational excursions into cryptography and miscellaneous inventions, some of which bordered on the eccentric. For example, he flirted with the idea of building a tick-tack-toe playing machine in order to raise funds for building the Analytical Engine.

One of the few bright spots that cheered Babbage's existence in the last twenty years of his life was the successful development of the Scheutz Difference Engine. In 1834 Dionysius Lardner, a well-known science lecturer and popularizer, had written an article in the *Edinburgh Review* describing Babbage's Difference Engine. This article was read by a Swedish printer and his son, Georg and Edvard Scheutz, who began to build a difference engine. After a heroic engineering effort of nearly twenty years that rivaled Babbage's own, they completed a full-scale engine, of which Babbage learned in 1852. The engine was exhibited at the Paris Exposition in 1855. Babbage visited the Exposition and was gratified to see the engine win the Gold Medal in its class. The following year it was purchased by the Dudley Observatory in Schenectady, New York, for $5,000. In Great Britain, the government provided £1,200 to have a copy made in order to compute a new set of life insurance tables.

Perhaps the most significant fact about the Scheutz Difference Engine was that it cost just £1,200—a fraction of what Babbage had spent. But inexpensiveness came at the cost of a lack of mechanical perfection; the Scheutz engine was not a particularly reliable machine and had to be adjusted quite frequently. Some other difference engines were built during the next half-century, but all of them were one-of-a-kind machines that failed to turn a profit for their inventors and never came near to establishing an industry. As a result, difference engines were largely abandoned and table-makers went back to using teams of human computers and conventional desk calculating machines.

Charles Babbage was never quite able to give up on his calculating engines, though. In 1856, after a gap of approximately ten years, he started work on the Analytical Engine again. Then in his mid-sixties, he realized that there was no serious prospect that it would ever be built. Right up to the end of his life he continued to work on the engine, almost entirely in isolation, filling notebook after notebook with impenetrable scribblings. He died on 18 October 1871, in his eightieth year, with no machine completed.

Why did Babbage fail to build any of his engines? Clearly his abrasive personality contributed to the failure, and for Babbage, the best was the enemy of the good. He also failed to understand that the British government cared not an iota about his calculating engine: The government was interested only in reducing the cost of table making, and it was immaterial whether the work was done by a calculating engine or by a team of human computers. But the primary cause of Babbage's failure was that he was pioneering in the 1820s and 1830s a digital approach to computing some fifty years before mechanical technologies had advanced sufficiently to make this relatively straightforward. If he had set out to do the same thing in the 1890s, the outcome might have been entirely different. As it was, Babbage's failure undoubtedly contributed to what L. J. Comrie called the "dark age" of digital computing. Rather than follow where Babbage had failed, scientists and engineers preferred to take a different, non-digital path—a path that involved the building of models, but which we now call analog computing.

The Tide Predictor and Other Analog Computing Machines

The adjective *analog* comes from the word *analogy*. The concept behind this mode of computing is that instead of computing with numbers, one builds a physical model, or analog, of the system to be investigated. A nice example of this approach was the orrery. This was a tabletop model of the planetary system named after the British Earl of Orrery, an aristocratic devotee of these gadgets in the early eighteenth century. An orrery consisted of a series of spheres. The largest sphere at the center of the orrery represented the sun; around the sun revolved a set of spheres representing the planets. The spheres were interconnected by a system of gear wheels, so that when a handle was turned, the planets would traverse their orbits according to the laws of planetary motion. In about ten minutes of handle turning, the Earth would complete a year-long orbit of the sun.

Orreries were much used for popular scientific lectures in Babbage's time, and many beautiful instruments were constructed for discerning gentlemen of learning and taste. Today they are highly valued museum pieces. Orreries were not much used for practical computing, however, because they were too inaccurate. Much later—in the 1920s—the principles were used in the planetariums seen in many science museums. In the planetarium an optical system projected the positions of the stars onto the hemispherical ceiling of a darkened lecture theater with high precision. For museum-goers of the 1920s and 1930s, the sight of the sky at night moving through the millennia was a remarkable and moving new experience.

The most important analog computing technology of the nineteenth century, from an economic perspective, was the mechanical tide predictor. Just as the navigation tables published in the *Nautical Almanac* helped minimize dangers at sea, so tide tables were needed to minimize the dangers of a ship being grounded when entering a harbor. Tides are caused primarily by the gravitational pull of the sun and the moon upon the Earth, but varying periodically in a very complex way as the sun and moon pursue their trajectories. Up to the 1870s the computations for tide tables were too time-consuming to be done for more than a few of the major ports of the world. In 1876 the British scientist Lord Kelvin invented a useful tide-predicting machine.

Lord Kelvin (who was William Thomson until his ennoblement) was arguably the greatest British scientist of his generation. His achievements occupy an astonishing nine pages in the British *Dictionary of National Biography*. He was famous to the general public as superintendent of the first transatlantic telegraph cable in 1866, but he also made many fundamental scientific contributions to physics, particularly radio telegraphy and wave mechanics. The tide predictor was just one of many inventions in his long life. It was an extraordinary-looking machine, consisting of a series of wires, pulleys, shafts, and gear wheels that simulated the gravitational forces on the sea and charted the level of water in a harbor as a continuous line on a paper roll. Tide tables could then be prepared by reading off the high and low points of the curve as it was traced.

Many copies of Kelvin's tide predictor were built to prepare tide tables for the thousands of ports around the world. Improved and much more accurate models continued to be made well into the 1950s, when they were finally replaced by digital computers. However, the tide predictor was a special-purpose instrument, not suitable for solving general scientific problems. This was the overriding problem with analog computing technologies—a specific problem required a specific machine.

The period between the two world wars was the heyday of analog computing. When a system could not be readily investigated mathematically, the best way forward was to build a model. And many wonderful models were built. For example, when a power dam needed to be constructed, the many equations that determined its behavior were far beyond the capability of the most powerful computing facilities then available. The solution was to build a scale model of the dam. It was then possible, for example, to measure the forces at the base of the dam and determine the height of waves caused by the wind blowing over the water's surface. The results could then be scaled up to guide the construction of the real dam. Similar techniques were used in land-reclamation planning in the Netherlands, where a fifty-foot scale model was constructed for tidal calculations of the Zuider Zee.

A whole new generation of analog computing machines was developed in the 1920s to help design the rapidly expanding U.S. electrical power system. During this decade, electricity supply expanded out of the metropolitan regions into the rural areas, especially in the south for textile manufacturing, and in California to power the land irrigation needed for the growth of farming. To meet this increased demand, the giant electrical machinery firms, General Electric and Westinghouse, began supplying AC generating equipment that was installed in huge power plants. These plants were then interconnected to form regional power networks.

The new power networks were not well understood mathematically; and when they were, the equations were too difficult to solve. Hence power network designers turned to the construction of analog models. These laboratory models consisted of circuits of resistors, capacitors, and inductors that emulated the electrical characteristics of the giant networks in the real world. Some of these "network analyzers" became fantastically complex. One of the most elaborate was the AC Network Calculator, built at MIT in 1930, which was 20 feet long and occupied an entire room. But like the tide predictor, the network analyzer was a special-purpose machine. Insofar as it became possible to expand the frontiers of analog computing, this was largely due to Vannevar Bush.

Vannevar Bush (1890–1974) is one of the key figures in twentieth-century American science, and although he was not a "computer scientist," his name will appear several times in this book. He first became interested in analog computing in 1912 when, as a student at Tufts College, he invented a machine he called a "profile tracer," which was used to plot ground contours. Bush described this machine in his autobiography *Pieces of the Action* in his characteristically folksy way:

> It consisted of an instrument box slung between two small bicycle wheels. The surveyor pushed it over a road, or across a field, and it automatically drew the profile as it went. It was sensitive and fairly accurate. If, going down a road, it ran over a manhole cover, it would duly plot the little bump. If one ran it around a field and came back to the starting point, it would show within a few inches the same elevation as that at which it had started.

Bush's homely style belied a formidable intellect and administrative ability. In some ways, his style was reminiscent of the fireside talks of President Roosevelt, of whom he was a confidant. In World War II, Bush became chief scientific adviser to Roosevelt, and director of the Office of Scientific Research and Development, which coordinated research for the war effort.

Tucked away in Bush's master's thesis of 1913 is a slightly blurred photograph taken a year earlier, showing a youthful and gangly Bush pushing his profile tracer across undulating turf. With its pair of small bicycle wheels and a black box filled with mechanical gadgetry, it captures perfectly the Rube Goldberg world of analog computing with its mixture of technology, gadgetry, eccentricity, inventiveness, and sheer panache.

In 1919 Bush became an instructor in electrical engineering at MIT, and the profile tracer concept lay at the core of the computing machines he went on to develop. In 1924 he had to solve a problem in electrical transmission, which took him and an assistant several months of painstaking arithmetic and graph plotting. "So some young chaps and I made a machine to do the work," Bush recalled. This machine was known as the "product integraph."

Like its predecessor, the product integraph was a one-problem machine. However, Bush's next machine, the "differential analyzer," was a powerful generalization of his earlier computing inventions that addressed not just a specific engineering problem but a whole class of engineering problems that could be specified in terms of ordinary differential equations. (These were an important class of equations that described many aspects of the physical environment involving rates of change, such as accelerating projectiles or oscillating electric currents.)

Bush's differential analyzer, built between 1928 and 1931, was not a general-purpose computer in the modern sense, but it addressed such a wide range of problems in science and engineering that it was far and away the single most important computing machine developed between the wars. Several copies of the MIT differential analyzer were built in the 1930s, at the University of Pennsylvania, at General Electric's main plant in Schenectady, New York, at the Aberdeen Proving Ground in Maryland, and abroad, at Manchester University in England and Oslo University in Norway. The possession of a differential analyzer made several of these places major centers of computing expertise during the early years of World War II—most notably the University of Pennsylvania, where the modern electronic computer was invented.

A Weather-Forecast Factory

Differential analyzers and network analyzers worked well for a limited class of engineering problems: those for which it was possible to construct a mechanical or electrical model of the physical system. But for other types of problems—table-making, astronomical calculations, or weather forecasting, for example—it was necessary to work with the numbers themselves. This was the digital approach.

Given the failure of Babbage's calculating engines, the only way of doing large-scale digital computations remained using teams of human computers. While the human computers would often work with logarithmic tables or calculating machines, the organizing principle was to use human beings to control the progress of the calculation.

One of the first advocates of the team computing approach was the Englishman Lewis Fry Richardson (1881–1953), who pioneered numerical meteorology—the application of numerical techniques to weather forecasting.

At that time weather forecasting was more an art than a science. Making a weather forecast involved plotting temperature and pressure contours on a weather chart and predicting—on the basis of experience and intuition—how the weather conditions would evolve. Richardson became convinced that forecasting tomorrow's weather on the basis of past experience—in the belief that history would repeat itself—was fundamentally unscientific. He believed that the only reliable predictor of the weather would be the mathematical analysis of the existing weather system.

In 1913 he was appointed director of the Eskdalemuir Observatory of the British Meteorological Office, in a remote part of Scotland, the perfect place to develop his theories of weather forecasting. For the next three years he wrote the first draft of his classic book, *Weather Forecasting by Numerical Process.* But Europe was at war and, although a pacifist, Richardson felt obliged to leave the safety of his distant Scottish outpost and join in the war effort by entering the Friends Ambulance Unit. Somehow, amid the carnage of war, he found the time in the intervals between transporting wounded soldiers from the front to test out his theories by calculating an actual weather forecast.

Weather prediction by Richardson's method involved an enormous amount of computation. As a test case, he took a set of known weather conditions for a day in 1910 and proceeded to make a weather forecast for six hours later, which he was then able to verify against the actual conditions recorded. The calculation took him six weeks, working in an office that was nothing more than a "heap of hay in a cold rest billet." He estimated that if he had thirty-two human computers, instead of just himself, he would have been able to keep pace with the weather. For technical reasons, the prediction was inaccurate but not a refutation of his ideas.

At the end of the war in 1918, Richardson returned to his work at the Meteorological Office and completed his book, which was published in 1922. In it he described one of the most extraordinary visions in the history of computing: a global weather-forecast factory. He envisaged having weather balloons spaced 200 kilometers apart around the globe that would take readings of wind speed, pressure, temperature, and other meteorological variables. This data would then be processed to make reliable weather forecasts. Richardson estimated that he would need 64,000 human computers to "race the weather for the whole globe." As he described his "fantasy" weather-forecast factory:

> Imagine a large hall like a theatre, except that the circles and galleries go right round through the space usually occupied by the stage. The walls of this chamber are painted to form a map of the globe. The ceiling represents the north polar regions, England is in the gallery, the tropics in the upper circle, Australia on the dress circle and the Antarctic in the pit. A myriad computers are at work upon the weather of the part of the map where each sits, but each computer attends only to one equation or part of an equation. The work of each region is coordinated by an official of higher rank. Numerous little "night signs" display the instantaneous values so that neighbouring computers can read them. Each number is thus displayed in three adjacent zones so as to maintain communication to the North and South on the map. From the floor of the pit a tall pillar rises to half the height of the hall. It carries a large pulpit on its top. In this sits the man in charge of the whole theatre; he is surrounded by several assistants and messengers. One of his duties is to maintain a uniform speed of progress in all parts of the globe. In this respect he is like the conductor of an orchestra in which the instruments are slide-rules and calculating machines. But instead of waving a baton he turns a beam of rosy light upon any region that is running ahead of the rest, and a beam of blue light upon those who are behind.

Richardson went on to explain how, as soon as forecasts had been made, they would be telegraphed by clerks to the four corners of the Earth. In the basement of the forecasting factory there would be a research department constantly at work trying to improve the numerical techniques. Outside, to complete his Elysian fantasy, Richardson imagined that the factory would be set amid "playing fields, houses, mountains and lakes, for it was thought that those who compute the weather should breathe of it freely."

When it was published in 1922, although Richardson's ideas about weather prediction were well received, his weather-forecasting factory fantasy was never taken seriously—and possibly he never intended that it should be. Nonetheless, he estimated that reliable weather forecasts would save the British economy £100 million a year, and even a weather-forecasting factory on the scale he proposed would have cost very much less than this to run.

But whether or not Richardson was serious, he never had the opportunity to pursue the idea. Even before the book was published, he had decided to withdraw from the world of weather forecasting because the British Meteorological Office had been taken over by the Air Ministry. As a Quaker, Richardson was unwilling to work directly for the armed services, so he resigned. He never worked on numerical meteorology again, but instead devoted himself to the mathematics of conflict and became a pioneer in peace studies. Because of the computational demands of numerical meteorology, it remained a backwater until the arrival of electronic computers after World War II, when both numerical meteorology and Richardson's reputation were resurrected. When Richardson died in 1953, it was clear that the science of numerical weather forecasting was going to succeed.

Scientific Computing Service

Richardson had designed his weather-forecasting factory around a team of human computers because it was the only practical digital computing technique he knew about, based on what he had learned of the working methods of the Nautical Almanac Office. In fact, computation for the *Nautical Almanac* had changed remarkably little since Babbage's time. The *Almanac* was still computed using logarithms by freelance computers, most of whom were retired members of staff. All this was soon to change, however, with the appointment of Leslie John Comrie (1893–1950), who would transform the approach to computing both in Britain and the United States.

Comrie did graduate work in astronomy at Cambridge University and then obtained a position at the Royal Observatory in Greenwich, where he developed a lifelong passion for computing. For centuries astronomers had made the heaviest use of scientific calculation of any of the sciences, so Comrie was in exactly the right environment for a person with his peculiar passion.

He was soon elected director of the computing section of the British Astronomical Association, where he organized a team of twenty-four human computers to produce astronomical tables. After completing his Ph.D. in 1923, he spent a couple of years in the United States, first at Swarthmore College and then at Northwestern University, where he taught some of the first courses in practical computing. In 1925 he returned to England to take up a permanent position with the Nautical Almanac Office.

Comrie was determined to revolutionize computing methods at the Nautical Almanac Office. Instead of using freelance computers, with their high degree of scientific training, he decided to systematize the work and make use of ordinary clerical labor and standard calculating machines. Almost all of Comrie's human computers were young, unmarried women with just a basic knowledge of commercial arithmetic.

Comrie's great insight was to realize that one did not need special-purpose machines such as differential analyzers; he thought computing was primarily a question of organization. For most calculations, he found that his "calculating girls" equipped with ordinary commercial calculating machines did the job perfectly well. Soon the Nautical Almanac Office was fitted out with Comptometers, Burroughs adding machines, and NCR accounting machines. Inside the Nautical Almanac Office, at first glance, the scene could easily have been mistaken for that in any ordinary commercial office. This was appropriate, for the Nautical Almanac Office was simply processing data that happened to be scientific rather than commercial.

Perhaps Comrie's major achievement at the Nautical Almanac Office was to bring punched-card machines—and therefore, indirectly, IBM—into the world of numerical computing. One of the most important and arduous computing jobs in the office was the annual production of a table of the positions of the moon, which was extensively used for navigation until very recent times. Calculating this table kept two human computers occupied year-round. Comrie showed considerable entrepreneurial flair in convincing his stuffy, government-funded employers to let him rent the punched-card machines, which were very costly compared with ordinary calculating machines. Before he could produce a single number, it was first necessary to transfer a large volume of tables, *Brown's Tables of the Moon*, onto half a million punched cards. Once this had been done, however, Comrie's punched-card machines churned out enough tables to satisfy the *Nautical Almanac* for the next twenty years in just seven months, and at a small fraction of the cost of computing them by hand. While Comrie was producing these tables he was visited by Professor E. W. Brown of Harvard University, the famous astronomer and the compiler of *Brown's Tables of the Moon*. Brown took away with him to the United States—and to IBM—the idea of using punched-card machines in scientific computation.

In 1930 Comrie's achievements were recognized by his promotion to director of the Nautical Almanac Office. From this position he became famous as the world's leading expert on numerical computing, and he was a consultant to many companies and scientific projects. However, after a few years he became increasingly frustrated with the bureaucracy of the Nautical Almanac Office, which resisted change and was unwilling to invest money to advance the state of the art of computing. In 1937 he took the unprecedented step of setting up in business as a commercial operation, the Scientific Computing Service Limited. It was the world's first for-profit calculating agency.

Comrie had perfect timing in launching his enterprise on the eve of World War II. He liked to boast that within three hours of Britain declaring war on Germany, he had secured a War Office contract to produce gunnery tables. He began with a staff of sixteen human computers, most of them young women. Comrie had a considerable flair for publicity and, as one headline in the popular magazine *Illustrated* put it, his "girls do the world's hardest sums."

Computer historians are now coming to realize that Comrie's work was just the tip of an iceberg of human computing during the war, which has hitherto been largely hidden. In the United States, particularly, we now know that there were several computing facilities that each made use of as many as a hundred human computers. The computers were mostly young female office workers called into the war effort, who had discovered in themselves a talent for computing. They were paid little more than the secretaries and typists from whose ranks they came; their work was "vital to the entire research effort, but remained largely invisible. Their names never appeared on the reports." When the war was over, these sisters of Rosie the Riveter returned to civilian life. Thus American society never had to confront the problem of what to do with the hundreds of human computers who would otherwise have soon been rendered obsolete by electronic computers.

The Harvard Mark I

The fact that there were so many human-computing organizations in existence is indicative of the demand for computing in the years up to World War II and during the war itself. In the period 1935–45, just as there was a boom in human-computing organizations, there was at the same time a boom in the invention of one-of-a-kind digital computing machines. There were at least ten machines constructed during this period, not only by government organizations but also by industrial research laboratories such as those of AT&T and RCA, as well as the technical departments of office-machine companies such as Remington

Rand, NCR, and IBM. Machines were built in Europe, too—in England and Germany.

These machines were typically used for table-making, ballistics calculations, and code-breaking. By far the best known, and the earliest, of these machines was the IBM Automatic Sequence Controlled Calculator. Known more commonly as the Harvard Mark I, this machine was constructed by IBM for Harvard University during 1937–43. The Harvard Mark I is important not least because it reveals how, as early as the 1930s, IBM was becoming aware of the convergence of calculating and office-machine technologies.

IBM's initial interest in digital computing dates from Professor Brown's visit to Comrie in the late 1920s. In 1929 Thomas Watson Sr., who was a benefactor of Columbia University, had endowed the university with a statistical laboratory and equipped it with standard IBM machines. One of Brown's most gifted students, Wallace J. Eckert (1902–71), joined the laboratory almost as soon as it was opened. Learning of Comrie's work from his former supervisor, Eckert persuaded IBM to donate more punched-card machines in order to establish a real computing bureau. Watson agreed to supply the machines for reasons that were partly altruistic, but also because he saw in Eckert and the Columbia University laboratory a valuable listening post. After the war Eckert and Columbia University were to play a key role in IBM's transition to computing.

Before the war, however, the chief beneficiary of IBM's computing largess was a Harvard researcher, Howard Hathaway Aiken (1900–1973). In 1936 Aiken was a graduate student in theoretical physics who was working on a problem concerning the design of vacuum tubes. In order to make progress with his dissertation, Aiken needed to solve a set of nonlinear differential equations. While analog machines such as Bush's Differential Analyzer at nearby MIT were suitable for ordinary differential equations, they could not generally solve nonlinear equations. It was this obstacle in the path of his research that inspired Aiken to put forward a proposal to the faculty of Harvard's physics department to construct a large-scale digital calculator. Not surprisingly, Aiken later recalled, his idea was received with "rather limited enthusiasm . . . if not downright antagonism."

Undaunted, Aiken took his proposal to George Chase, the chief engineer of the Monroe Calculating Machine Company, one of the leading manufacturers of adding and calculating machines. Chase was a very well-known and respected figure in the calculating-machine industry, and he urged his company to sponsor Aiken's machine "in spite of the large anticipated costs because he believed the experience of developing such a sophisticated calculator would be invaluable to the company's future." Unfortunately for Chase and Aiken (and in the long run, for Monroe), the company turned the proposal down. Chase, however, remained convinced of the importance of the concept and urged Aiken to approach IBM.

In the meantime, word of Aiken's calculating-machine proposal had begun to spread in the Harvard physics laboratory, and one of the technicians approached him. Aiken recalled being told by the technician that he "couldn't see why in the world I wanted to do anything like this in the Physics Laboratory, because we already had such a machine and nobody ever used it." Puzzled, Aiken allowed the technician to take him up to the attic, where he was shown a fragment of Babbage's calculating engine set on a wooden stand about 18 inches square.

The engine fragment had been donated to Harvard in 1886 by Babbage's son, Henry. Henry had scrapped the remains of his father's engines, but not quite everything had been consigned to the melting pot. He held on to a number of sets of calculating wheels and had them mounted on wooden display stands. Just one set was sent abroad—to the president of Harvard College. In 1886 the college was celebrating its 250th anniversary, and Henry likely thought that it was the most fertile place in which he could plant the seed of Babbage in the New World.

This was the first time that Aiken had ever heard of Babbage, and it was probably the first time the calculating wheels had seen the light of day in decades. Aiken was captivated by the artifact. Of course, the Babbage computing wheels were no more than a fragment. They were not remotely capable of computation, and they were built in a technology that had long been superseded by the modern calculating machine. But for Aiken they were a relay baton carried from the past, and he consciously saw himself as Babbage's twentieth-century successor. To find out more about Babbage, Aiken immediately went to the Harvard University library, where he obtained a copy of Babbage's autobiography, *Passages from the Life of a Philosopher*, published in 1864. In the pages of that book, Aiken read:

> If, unwarned by my example, any man shall undertake and shall succeed in really constructing an engine embodying in itself the whole of the executive department of mathematical analysis upon different principles or by simpler mechanical means, I have no fear of leaving my reputation in his charge, for he alone will be fully able to appreciate the nature of my efforts and the value of their results.

When Aiken came across this passage, he "felt that Babbage was addressing him personally from the past."

In order to gain sponsorship for his calculator from IBM, Aiken secured an introduction to James Bryce, IBM's chief engineer. Bryce immediately understood the importance of Aiken's proposal. Fortunately for Aiken, Bryce enjoyed much greater confidence from his employers than did Chase at Monroe. Bryce had no trouble convincing Watson to back the project, initially to the tune of $15,000. The following month, December 1937, Aiken polished up his proposal, in which he referred to some of Babbage's concepts—in particular, the idea of using Jacquard cards to program the machine—and sent it to IBM. This proposal is the earliest surviving document about the Mark I, and it must remain something of an open question as to how much Aiken derived from Charles Babbage. In terms of design and technology, the answer is clearly very little; but in terms of a sense of destiny, Aiken probably derived a great deal more.

At this point Aiken's proposal barely constituted a computer design. Indeed, it really amounted to a general set of functional requirements. He had specified that the machine should be "fully automatic in its operation once a process is established" and that it should have a "control to route the flow of numbers through the machine." In effect, Aiken had sketched the layout of a house, and now it was left to IBM to draw up the detailed plans and build it.

At Bryce's suggestion, Aiken attended an IBM training school, where he learned how to use IBM machines and obtained a thorough understanding of the capabilities and limitations of the technology. Then, early in 1938, he visited IBM's Endicott Laboratory in New York, where he began to work with a team of Bryce's most trusted and senior engineers. Bryce placed Claire D. Lake in charge of the project. Lake was one of IBM's old-time engineer inventors and the designer of IBM's first printing tabulator in 1919. He in turn was supported by two of IBM's very best engineers.

The initial estimate of $15,000 to build the machine was quickly raised to $100,000, and early in 1939 the IBM board gave formal approval for the construction of an "Automatic Computing Plant." Aiken's role was now essentially that of a consultant and, other than spending parts of the summers of 1939 and 1940 with IBM, he took little part in the detailed design and engineering of the machine. By 1941 Aiken had been enlisted by the navy, and the calculating machine had become just one of many special development projects for the armed forces within IBM. There were many delays, and it was not until January 1943 that the machine ran its first test problem.

The 5-ton calculator was a monumental piece of engineering. Because all the basic calculating units had to run in mechanical synchronism, they were all lined up and driven by a 50-foot shaft, powered by a 5-horsepower electric motor, like "a nineteenth-century New England textile mill." The machine was thus fifty-one feet wide but only two feet deep. All together it contained three-quarters of a million parts and had hundreds of miles of wiring. In operation, the machine was described by one commentator as making the sound of "a roomful of ladies knitting." The machine could store just seventy-two numbers, and was capable of three additions or subtractions a second. Multiplication took six seconds, while calculating a logarithm or a trigonometrical function took over a minute.

Even by the standards of the day, these were pedestrian speeds. The significance of the Mark I lay not in its speed but in the fact that it was the first fully automatic machine to be completed. Once set in operation, the machine would run for hours or even days doing what Aiken liked to call "makin' numbers." The machine was programmed by a length of paper tape some three inches wide on which "operation codes" were punched. Most programs were repetitive in nature—for example, computing the successive values of a mathematical table—so that the two ends of the tape could be glued together to form a loop. Short programs might contain a hundred or so operations, and the whole loop would rotate about once a minute. Thus the machine would take perhaps half a day to calculate one or two pages of a volume of mathematical tables.

Unfortunately, the calculator was incapable of making what we now call a "conditional branch"—that is, changing the progress of a program according to the results of an earlier computation. This made complex programs physically very long. So that the program tapes did not get soiled on the floor, special racks had to be built onto which a program tape could be stretched over pulleys to take up the slack. To run such a program, the rack had to be trundled from a storeroom and docked onto the machine. If Aiken had studied Babbage's—and especially Lovelace's—writings more closely, he would have discovered that Babbage had already arrived at the concept of a conditional branch. In this respect, the Harvard Mark I was much less impressive than Babbage's Analytical Engine, designed a century earlier.

The Mark I began to run test problems in secret in early 1943, but was not used on production work until it was moved to the Harvard campus a year later. Even then the machine was not used heavily until Aiken, still an officer in the Naval Warfare School, was recalled to Harvard to take charge of the machine. Aiken liked to say, "I guess I'm the only man in the world who was ever commanding officer of a computer." Serious computing began for the Bureau of Ships in May 1944, mainly producing mathematical tables.

Back at Harvard, Aiken planned a ceremony to inaugurate the calculator in collaboration with IBM. At this stage the calculator looked very much like what it was: a long row of naked electromechanical punched-card equipment. Aiken preferred the machinery this way so that it was easy to maintain and modify the calculator. Thomas Watson, however, was anxious to secure the maximum publicity for the machine and insisted that it be fitted out in suitable IBM livery. Overruling Aiken, he commissioned the industrial designer Norman Bel Geddes to give it a more attractive face. Geddes outdid himself, producing a gleaming creation in stainless steel and polished glass. It was the very personification of a "giant brain."

The IBM Automatic Sequence Controlled Calculator was officially dedicated at Harvard on 7 August 1944, some seven years after Aiken's first contact with IBM. At the ceremony, Aiken's arrogance overcame his judgment to a degree that his reputation never fully recovered. First, he took credit as the sole inventor of the calculator and refused to recognize the contribution made by IBM's engineers, who had spent years making his original proposal concrete. Even worse, from Watson's point of view, he failed to acknowledge the fact that IBM had underwritten the entire project. According to his biographers, Watson was pale and shaken with anger: "Few events in Watson's life infuriated him as much as the shadow cast on his company's achievement by that young mathematician. In time his fury cooled to resentment and a desire for revenge, a desire that did IBM good because it gave him an incentive to build something better in order to recapture the spotlight." (That "something better" turned out to be a machine known as the Selective Sequence Electronic Calculator, is discussed in chapter 5.)

The dedication of the Harvard Mark I captured the imagination of the public to an extraordinary extent and gave headline writers a field day. *American Weekly* called it "Harvard's Robot Super-Brain," while *Popular Science Monthly* declared "Robot Mathematician Knows All the Answers." After the dedication and the press coverage, there was intense interest in the Harvard Mark I from scientific workers and engineers wanting to use it. This prompted Aiken and his staff to produce a 500-page *Manual of Operation of the Automatic Sequence Controlled Calculator*, which was effectively the first book on digital computing ever published. Back in England, Comrie wrote a review of it in *Nature*, in which he made his celebrated remark about the black mark earned by the British government for failing to construct Babbage's original calculating engine. If Comrie remained annoyed with the British government of the previous century, he was at least as annoyed at Aiken's lack of grace in not acknowledging IBM's contribution: "One notes with astonishment, however, the significant omission of 'IBM' in the title and in Prof. Aiken's preface." Comrie well understood the role played by IBM and its engineers in making Aiken's original proposal a reality.

Although the Harvard Mark I was a milestone in the history of computing, it had a short heyday before being eclipsed by electronic machines, which operated much faster because they had no moving parts. This was true of virtually all the electromechanical computers built in the 1930s and 1940s. Above all the Harvard Mark I was profoundly slow. While Aiken liked to claim that it was a hundred times faster than a human computer and could perform six man-months of computing in a day, it was capable of a mere 200 operations per minute.

Not only was the Harvard Mark I a technological dead end, it did not even do anything very useful in the fifteen years that it ran. It was used in a number of applications for the navy, and here the machine was sufficiently useful that the navy commissioned additional computing machines from Aiken's laboratory. Aiken, like many other scientists of his time, had great hopes for the scientific value of the machine in producing mathematical tables for use in scientific research. The Harvard Mark I was used to produce twenty-five volumes of mathematical tables, but like other tables produced by early computers, these had a limited use and a short life. Scientific researchers found that they could usually have the computer produce these tabular values on demand, and there was no reason to calculate them and print them out in tables.

The Harvard Mark I, however, was a fertile training ground for early computer pioneers, such as Grace Murray Hopper, and Aiken had great influence on the design of early computing machines in Europe, where electronics were not yet readily available in the war-torn communities. But the real importance of the Harvard Mark I was as an icon of the computer age—the first fully automatic computer to come into operation. The significance of this event was widely appreciated by scientific commentators, and the machine also had an emotional appeal as a final vindication of Babbage's life. In 1864 Babbage had written: "Half a century may probably elapse before anyone without those aids which I leave behind me, will attempt so unpromising a task." Even Babbage had underestimated just how long it would take.

In terms of technology and speed, the Harvard Mark I was fairly described as "Babbage's dream come true." But if Babbage had been around to see it, he would not have been unduly impressed. In its technological capabilities, it was very much the kind of computer he had proposed in the 1830s. This was particularly true with regard to speed. It was perhaps only ten times faster than the speed he had planned for the Analytical Engine. Babbage could never have envisioned that one day electronic machines would come onto the scene with speeds *thousands* of times faster than he ever dreamed. This happened within two years of the completion of the Harvard Mark I.

Part Two

CREATING THE COMPUTER

4

:: INVENTING THE COMPUTER

WORLD WAR II was a scientific war: Its outcome was determined largely by the effective deployment of scientific research and technical developments. The best-known wartime scientific program was the Manhattan Project at Los Alamos to develop the atomic bomb. Another major program of the same scale and importance as atomic energy was radar, in which the Radiation Laboratory at MIT played a major role. It has been said that although the bomb ended the war, radar won it.

Emphasis on these major programs can overshadow the rich tapestry of the scientific war effort. One of the threads running through this tapestry was the need for mathematical computation. For the atomic bomb, for example, massive computations had to be performed to perfect the explosive lens that assembled a critical mass of enriched uranium. At the outbreak of war, the only computing technologies available were analog machines such as differential analyzers, primitive digital technologies such as punched-card installations, and teams of human computers equipped with desktop calculating machines. Even relatively slow one-of-a-kind mechanical computers such as the Harvard Mark I lay some years in the future.

The scientific war effort in the United States was administered by the Office of Scientific Research and Development (OSRD). This organization was headed by Vannevar Bush, the former professor of electrical engineering at MIT and inventor of the Differential Analyzer. Bush was a brilliantly effective research director. Although he had developed analog computing machines in the 1930s, by the outbreak of war he had ceased to have any active interest in computing, despite his understanding of its importance in scientific research.

As it happened, the invention of the modern electronic computer occurred in an obscure laboratory, and it was not until very late in the war that it even came to Bush's notice. Thus, Bush's participation was entirely administrative, and one of his key contributions was to propose an Act of Congress by which the research laboratories of the Navy and War departments were united with the hundreds of civil research establishments under the National Defense Research Committee (NDRC), which would "coordinate, supervise, and conduct scientific research on the problems underlying the development, production, and use of mechanisms and devices of warfare, except scientific research on the problems of flight."

The NDRC consisted of twelve leading members of the scientific community who had the authority and confidence to get business done. Bush allowed research projects to percolate up from the bottom as well as being directed from the top. He later wrote that "most of the worthwhile programs of NDRC originated at the grass roots, in the sections where civilians who had specialized intensely met with military officers who knew the problem in the field." Once a project had been selected, the NDRC strove to bring it to fruition; as Bush recalled:

> When once a project got batted into [a] form which the section would approve, with the object clearly defined, the research men selected, a location found where they could best work, and so on, prompt action followed. Within a week NDRC could review the project. The next day the director could authorize, the business office could send out a letter of intent, and the actual work could start. In fact it often got going, especially when a group program was supplemented, before there was any formal authorization.

This description conveys well the way the first successful electronic computer in the United States grew out of an obscure project in the Moore School of Electrical Engineering at the University of Pennsylvania.

The Moore School

Computing was not performed for its own sake, but always as a means to an end. In the case of the Moore School, the end in view was ballistics computations for the Aberdeen Proving Ground in Maryland, which was responsible for commissioning armaments for the U.S. Army. The proving ground was a large tract of land on the Chesapeake Bay, where newly developed artillery and munitions could be test-fired and calibrated. The Aberdeen Proving Ground lay about 60 miles to the southwest of Philadelphia, and the journey between it and the Moore School was easily covered by railroad in a little over an hour.

In 1935, long before the outbreak of war, a research division was established at the proving ground, which acquired one of the first copies of the Bush Differential Analyzer. The laboratory had been founded during World War I and was formally named the Ballistics Research Laboratory (BRL) in 1938. Its main activity was mathematical ballistics—calculating the trajectories of weapons fired through the air or water—and it initially had a staff of about thirty people. When hostilities began in Europe in the 1930s, the scale of operations at the BRL escalated. When the United States entered the war, top-flight scientists in mathematics, physics, astrophysics, astronomy, and physical chemistry were brought in, as well as a much larger number of enlisted junior scientists who assisted them. Among the latter was a young mathematician named Herman H. Goldstine, who was later to play an important role in the development of the modern computer.

What brought the BRL and the Moore School of Electrical Engineering together was the fact that the latter also had a Bush Differential Analyzer. The BRL used the Moore School's differential analyzer for its growing backlog of ballistics calculations and, as a result, established a close working relationship with the staff. Although the Moore School was one of the better-established electrical engineering schools in the United States, it had nowhere near the prestige or resources of MIT, which received the bulk of the wartime research projects in electronics.

In the months preceding World War II, the Moore School was placed on a war footing: The undergraduate programs were accelerated by eliminating vacations, and the school took on war training and electronics-research programs. The main training activity was the Engineering, Science, Management, War Training (ESMWT) program, which was an intensive ten-week course designed to train physicists and mathematicians for technical posts—particularly in electronics, in which there was a great manpower shortage. Two of the outstanding graduates of the ESMWT program in the summer of 1941 were John W. Mauchly, a physics instructor from Ursinus College, a small liberal arts college not far from Philadelphia, who had an interest in numerical weather prediction, and Arthur W. Burks, a mathematically inclined philosopher from the University of Michigan. Instead of moving on to technical posts, they accepted invitations to stay on as instructors at the Moore School, where they were to become two of the key players in the invention of the modern electronic computer. Another training program at the Moore School was organized for the

BRL to train female "computers" to operate desk calculating machines. This was the first time women had been admitted to the Moore School. An undergraduate at the time recalled:

> Mrs. John W. Mauchly, a professor's wife, had the job of instructing the women. Two classrooms were taken over for their training, and each girl was given a desk calculator. The girls sat at their desks for hours, pounding away at the calculators until the clickety clack merged into a symphony of metallic digits. After the girls were trained, they were put to work calculating ballistic missile firing tables. Two classrooms full of the girls doing just that, day after day, was impressive. The groundwork was being established for the invention of the computer.

Eventually the hand-computing staff consisted of about two hundred women.

By early 1942, the Moore School was humming with computing activity. Down in the basement of the building, the differential analyzer—affectionately known as "Annie"—was in constant use. The same undergraduate recalled that "on the very few occasions we were allowed to see Annie, we were amazed by the tremendous maze of shafts, gears, motors, servos, etc." In the same building, a hundred women computers were engaged in the same work, using desk calculating machines.

All this computing activity was needed to produce "firing tables" for newly developed artillery, and for older equipment being used in new theaters of war. As any novice rifleman soon discovers, one does not hit a distant target by firing directly at it. Rather, one fires slightly above it, so that the bullet flies in a parabolic trajectory—first rising into the air, reaching its maximum height midway, then falling toward the target. In the case of a World War II weapon, with a range of a mile or so, one could not possibly use guesswork and rule of thumb to aim the gun; it would have taken dozens of rounds to find the range. There were simply too many variables to take into account. The flight of the shell would be affected by the head- and cross-winds, the type of shell, the air temperature, and even local gravitational conditions. For most weapons, the gunner was provided with a firing table in the form of a pocket-sized booklet. Given the range of a target, the firing table would enable the gunner to determine the angle at which to fire the gun into the air (the elevation) and the angle to the target (the azimuth). More sophisticated weapons used a "director," which automatically aimed the gun using data entered by the gunner. Whether the gun was aimed manually or using a director, the same kind of firing table had to be computed.

A typical firing table contained data for around 3,000 trajectories. To calculate a single one of these required the mathematical integration of an ordinary differential equation in seven variables, which took the differential analyzer ten to twenty minutes. Working flat out, a complete table would take about thirty days. Similarly, a "calculating girl" at a desk machine would take one or two days to calculate a trajectory, so that a 3,000-entry firing table would tie up a hundred-strong calculating team for perhaps a month. The lack of an effective calculating technology was thus a major bottleneck to the effective deployment of the multitude of newly developed weapons.

The Atanasoff-Berry Computer

In the summer of 1942 John Mauchly proposed the construction of an electronic computer to eliminate this bottleneck. Although he was an assistant professor with teaching responsibilities and did not have any formal involvement with the computing activity, he was married to a woman, Mary, who was an instructor of the women computers, and so was well aware of the sea of numbers in which the organization was beginning to drown. Moreover, he had a knowledge of computing because of his interest in numerical weather forecasting, a research interest that long predated his arrival at the Moore School.

Mauchly's was by no means the only proposal for an electronic computer to appear during the war. For example, Konrad Zuse in Germany was already secretly in the process of constructing a machine; and the code-breaking establishments of both the United States and Britain embarked on electronic computer projects at about the same time as the Moore School. But it would not be until the 1970s that the general public would learn anything about these other projects; they had little impact on the development of the modern computer, and none at all on the Moore School's work. There was, however, another computer project at Iowa State University that was to play an indirect part in the invention of the computer at the Moore School. Although the Iowa State project was not secret, it was obscure, and little was known of it until the late 1960s.

The project was begun in 1937 by John Vincent Atanasoff, a professor of mathematics and physics. He enlisted the help of a graduate student, Clifford Berry, and by 1939 they had built a rudimentary electronic computing device. During the next two years they went on to build a complete machine—later known as the Atanasoff-Berry Computer (ABC). They developed several ideas that were later rediscovered in connection with electronic computers, such as binary arithmetic and electronic switching elements. In December 1940, while Mauchly was still an instructor at Ursinus College, Atanasoff attended a lecture Mauchly presented to the American Association for the Advancement of Science in which he described a primitive analog computing machine he had devised for weather prediction. Atanasoff, discovering a like-minded scientist, introduced himself and invited Mauchly to visit to see the ABC. Mauchly crossed the country to visit Atanasoff the following June, shortly before enrolling at the Moore School. He stayed for five days in Atanasoff's home and learned everything he could about the machine. They parted on very amicable terms.

Atanasoff, described by his biographer as the "forgotten father of the computer," occupies a curious place in the history of computing. According to some writers, and some legal opinions, he is the true begetter of the electronic computer; and the extent to which Mauchly drew on Atanasoff's work would be a key issue in determining the validity of the early computer patents fifteen years later. There is certainly no question that Mauchly visited Atanasoff in June 1941 and received a detailed understanding of the ABC, although Mauchly subsequently stated that he took "no ideas whatsoever" from Atanasoff's work. As for the ABC itself, in 1942, before the computer could be brought into reliable operation, Atanasoff was recruited for war research in the Naval Ordnance Laboratory. A brief effort to build computers for the navy failed, and he never resumed active research in computing after the war. It was only in the late 1960s that his early contributions became widely known.

The extent to which Mauchly drew on Atanasoff's ideas remains unknown, and the evidence is massive and conflicting. The ABC was quite modest technology, and it was not fully implemented. At the very least we can infer that Mauchly saw the potential significance of the ABC and that this may have led him to propose a similar, electronic solution to the BRL's computing problems.

Eckert and Mauchly

Mauchly, with a kind of missionary zeal, discussed his ideas for an electronic computing machine with any of his colleagues at the Moore School who would listen. The person most taken with these ideas was a young electronic engineer named John Presper Eckert. A twenty-two-year-old research associate who had only recently completed his master's degree, Eckert was "undoubtedly the best electronic engineer in the Moore School." He and Mauchly had already established a strong rapport by working on the differential analyzer, in which they had replaced some of the mechanical integrators with electronic amplifiers, getting it "to work about ten times faster and about ten times more accurately than it had worked before." In this work they frequently exchanged ideas about electronic computing—Mauchly invariably in the role of a conceptualizer and Eckert in that of a feet-on-the-ground engineer. Eckert immersed himself in the literature of pulse electronics and counting circuits and rapidly became an expert.

The Moore School, besides its computing work for BRL, ran a number of research and development projects for the government. One of the projects, with which Eckert was associated, was for something called a delay-line storage system. This was completely unconnected with computing but would later prove to be a key enabling technology for the modern computer. The project had been subcontracted by MIT's Radiation Laboratory, which required some experimental work on delay-line storage for a Moving Target Indicator (MTI) device. The MTI device was a critical problem in the newly developed radar systems. Although radar enabled an operator to "see in the dark" on a cathode-ray tube display, it suffered from the problem that it "saw" everything in its range—not just military objects but also land, water, buildings, and so on. The result was that the radar screen was cluttered with background information that made it difficult to discern military objectives. The aim of the MTI device was to discriminate between moving and stationary objects by canceling out the radar trace of the latter; this would remove much of the background clutter, allowing moving targets to stand out clearly.

The MTI device worked as follows. As a radar echo was received, it was stored and subtracted from the next incoming signal; this canceled out the radar signal reflected by objects whose position had not changed between consecutive bursts of radar energy. Moving objects, however, would not be canceled out since their position would have changed between one radar burst and the next; they would thus stand out on the display. To achieve this it was necessary to store an electrical signal for a thousandth of a second or so. This was where the delay line came in. It made use of the fact that the velocity of sound is only a tiny fraction of the velocity of light. To achieve storage, the electrical signal was converted to sound, passed through an acoustic medium, and converted back to an electrical signal at the other end—typically about a millisecond later. Various fluids had been tried, and at this time Eckert was experimenting with a mercury-filled steel tube.

By August 1942 Mauchly's ideas on electronic computing had sufficiently crystallized that he wrote a memorandum on *The Use of High Speed Vacuum Tubes for Calculating*. In it he proposed an "electronic computor" that would be able to perform calculations in 100 seconds that would take a mechanical differential analyzer 15 to 30 minutes, and that would have taken a human computer "*at least* several hours." This memorandum was the real starting point for the electronic computer project. Mauchly submitted it to the director of research of the Moore School and to the Army Ordnance department, but it was ignored by both.

The person responsible for getting Mauchly's proposal taken seriously was Lieutenant Herman H. Goldstine. A mathematics Ph.D. from the University of Chicago, Goldstine had joined the BRL in July 1942 and was initially assigned as the liaison officer with the Moore School on the project to train female computers. He had not received Mauchly's original memo, but during the fall of 1942 he and Mauchly had fairly frequent conversations on the subject of electronic computing. During this time the computation problem at BRL had begun to reach a crisis, and Goldstine reported in a memo to his commanding officer:

> In addition to a staff of 176 computers in the Computing Branch, the Laboratory has a 10-integrator differential analyzer at Aberdeen and a 14-integrator one at Philadelphia, as well as a large group of IBM machines. Even with the personnel and equipment now available, it takes about three months of work on a two-shift basis to turn out the data needed to construct a director, gun sight, or firing table.

ENIAC and EDVAC:
The Stored-Program Concept

By the spring of 1943 Goldstine had become convinced that Mauchly's proposal should be resubmitted, and told Mauchly that he would use what influence he could to promote it. By this time the original copies of Mauchly's report had been lost, so a fresh version was retyped from his notes. This document, dated 2 April 1943, was to serve as the basis of a contract between the Moore School and BRL. Things now moved ahead very quickly. Goldstine had prepared the ground carefully, organizing a number of meetings in the BRL during which the requirements of an electronic calculator were hammered out. This resulted in Eckert and Mauchly's original plan for a machine containing 5,000 tubes and costing $150,000 to escalate to one containing 18,000 tubes and costing $400,000. On 9 April a formal meeting was arranged with Eckert and Mauchly, the research director of the Moore School, Goldstine, and the director of the BRL. At this meeting the terms of a contract between the two organizations were finalized.

That same day, Project PX to construct this machine, known as the ENIAC (Electronic Numerical Integrator and Computer), began. It was Presper Eckert's twenty-fourth birthday, and the successful engineering of the ENIAC was to be very much his achievement.

The ENIAC was not without its critics. Indeed, within the NDRC the attitude was "at best unenthusiastic and at worst hostile." The most widely voiced criticism against the project concerned the large number of vacuum tubes the machine would contain—estimated at around 18,000. It was well known that the life expectancy of a tube was around 3,000 hours, which meant, by a naive calculation, that there would be a tube failure every ten minutes. This estimate did not take account of the thousands of resistors, capacitors, and wiring terminals—all of which were potential sources of breakdown. Eckert's critical engineering insight in building the ENIAC was to realize that, although the mean life of electronic tubes was 3,000 hours, this was only when a tube was run at its full operating voltage. If the voltage was reduced to, say, two-thirds of the nominal value, then its life was extended to tens of thousands of hours. Another insight was his realization that most tube failures occurred when tubes were switched on and warmed up. For this reason radio transmitters of the period commonly never switched off the "heaters" of their electronic tubes. Eckert proposed to do the same for the ENIAC.

The ENIAC was many times more complex than any previous electronic system, in which a thousand tubes would have been considered unusual. All together, in addition to its 18,000 tubes, the ENIAC would include 70,000 resistors, 10,000 capacitors, 6,000 switches, and 1,500 relays. Eckert applied rigorous testing procedures to all the components. Whole boxes of resistors were brought in and their tolerances were checked one by one on a specially designed test rig. The best resistors were kept to one side for the most critical parts of the machine while others were used in less sensitive parts. On other occasions whole boxes of components would be rejected and shipped back to the manufacturer.

The ENIAC was Eckert and Mauchly's project. Eckert, with his brittle and occasionally irascible personality, was "the consummate engineer," while Mauchly, always quiet, academic, and laid back, was "the visionary." The physical construction of the ENIAC was undoubtedly Eckert's achievement. He showed an extraordinary confidence for a young engineer in his mid-twenties:

> Eckert's standards were the highest, his energies almost limitless, his ingenuity remarkable, and his intelligence extraordinary. From start to finish it was he who gave the project its integrity and ensured its success. This is of course not to say that the ENIAC development was a one man show. It was most

clearly not. But it was Eckert's omnipresence that drove everything forward at whatever cost to humans including himself.

Eckert recruited to the project the best graduating Moore School students, and by the autumn of 1943 he had built up a team of a dozen young engineers. In a large back room of the Moore School, the individual units of the ENIAC began to take shape. Engineers were each assigned a space at a worktable that ran around the room, while assemblers and wiremen occupied the central floor space.

After the construction of the ENIAC had been going on for several months, it became clear that there were some serious defects in the design. Perhaps the most serious problem was the time it would take to reprogram the machine when one problem was completed and another was to be run. Machines such as the Harvard Mark I were programmed using punched cards or paper tape, so that the program could be changed as easily as fitting a new roll onto a player piano. This was possible only because the Harvard machine was electro-mechanical and its speed (three operations per second) was well matched to the speed at which instructions could be read from a paper tape. But the ENIAC would perform 5,000 operations per second, and it was infeasible to use a card or tape reader at that kind of speed. Instead, Eckert and Mauchly decided the machine would have to be specially wired-up for a specific problem. At first glance, when programmed, the ENIAC would look rather like a telephone exchange, with hundreds of patch cords connecting up the different units of the machine to route electrical signals from one point to another. It would take hours, perhaps days, to change the program—but Eckert and Mauchly could see no alternative.

During the early summer of 1944, when the ENIAC had been under construction for about eighteen months, Goldstine had a chance encounter with John von Neumann. Von Neumann was the youngest member of the Institute for Advanced Study in Princeton, where he was a colleague of Einstein and other eminent mathematicians and physicists. Unlike most of the other people involved in the early development of computers, von Neumann had already established a world-class scientific reputation (for providing the mathematical foundations of quantum mechanics and other mathematical studies). Despite being from a wealthy banking family, von Neumann had suffered under the Hungarian government's persecution of Jews in the early 1930s. He had a hatred for totalitarian governments and readily volunteered to serve as a civilian scientist in the war effort. His formidable analytical mind and administrative talents were quickly recognized, and he was appointed a consultant to several wartime projects. He had been making a routine consulting trip to BRL when he met Goldstine. Goldstine, who had only attained the rank of assistant professor in civilian life, was somewhat awestruck in the presence of the famous mathematician. He recalls:

> I was waiting for a train to Philadelphia on the railroad platform in Aberdeen when along came von Neumann. Prior to that time I had never met this great mathematician, but I knew much about him of course and had heard him lecture on several occasions. It was therefore with considerable temerity that I approached this world-famous figure, introduced myself, and started talking. Fortunately for me von Neumann was a warm, friendly person who did his best to make people feel relaxed in his presence. The conversation soon turned to my work. When it became clear to von Neumann that I was concerned with the development of an electronic computer capable of 333 multiplications per second, the whole atmosphere of our conversation changed from one of relaxed good humor to one more like the oral examination for the doctor's degree in mathematics.

At this time, von Neumann was a consultant to the Manhattan Project to build the atomic bomb. This project was, of course, highly secret and was unknown to anyone at Goldstine's level in the scientific administration. Von Neumann had become a consultant to Los Alamos in late 1943 and was working on the mathematics of implosions. In the atomic bomb it was necessary to force two hemispheres of enriched uranium together, uniformly from all sides, to form the critical mass that gave rise to the nuclear explosion. To prevent premature detonation, the two hemispheres had to be forced together in a few thousandths of a second using an explosive charge surrounding them. The mathematics of this implosion problem, which involved the solution of a large system of partial differential equations, was very much on von Neumann's mind when he met Goldstine that day.

How was it that von Neumann was unaware of the ENIAC, which promised to be the most important computing development of the war? The simple answer is that no one at the top, that is, in the NDRC, had sufficient faith in the project to bother bringing it to von Neumann's attention. Indeed, while he had been wrestling with the mathematics of implosion, von Neumann had written in January 1944 to Warren Weaver, head of the OSRD applied mathematics panel and a member of the NDRC, inquiring about existing computing facilities. He had been told about the Harvard Mark I, about work going on at the Bell Telephone Laboratories, and about IBM's war-related computing projects. But the ENIAC was never mentioned: Weaver apparently did not want to waste von Neumann's time on a project he judged would never amount to anything.

However, now that von Neumann had discovered the ENIAC he wanted to learn more, and Goldstine arranged for him to visit the Moore School in early August. The whole group was in awe of von Neumann's reputation, and there was great expectation of what he could bring to the project. Goldstine recalls that Eckert had decided he would be able to tell whether von Neumann was really a genius by his first question: If it turned out to be about the logical structure of the machine, then he would be convinced. Goldstine recalled, "Of course, this *was* von Neumann's first query."

When von Neumann arrived, the construction of the ENIAC was in high gear. The Moore School's technicians were working two shifts, from 8:30 in the morning until 12:30 at night. The design of the ENIAC had been frozen two months previously, and, as had its designers, von Neumann soon became aware of its shortcomings. He could see that while the machine would be useful for solving the ordinary differential equations used in ballistics computations, it would be far less suitable for solving his partial differential equations—largely because the total storage capacity of twenty numbers was far too little. He would need storage for hundreds if not thousands of numbers. Yet the tiny amount of storage that was provided in the ENIAC accounted for over half of its 18,000 tubes. Another major problem with the ENIAC, which was already obvious, was the highly inconvenient and laborious method of programming, which involved replugging the entire machine. In short, there were three major shortcomings of the ENIAC: too little storage, too many tubes, too lengthy reprogramming.

Von Neumann was captivated by the logical and mathematical issues in computer design and became a consultant to the ENIAC group to try to help them resolve the machine's deficiencies and develop a new design. This new design would become known as the "stored-program computer," on which virtually all computers up to the present day have been based. All this happened in a very short space of time.

Von Neumann's arrival gave the group the confidence to submit a proposal to the BRL to develop a new, post-ENIAC machine. It helped that von Neumann was able to attend the BRL board meetings where the proposal was discussed. Approval for "Project PY" was soon forthcoming, and funding of $105,000 was provided. From this point on, while construction of the ENIAC

continued, all the important intellectual activity of the computer group revolved around the design of ENIAC's successor: the EDVAC, for Electronic Discrete Variable Automatic Computer.

The shortcomings of the ENIAC were very closely related to its limited storage. Once that had been solved, most of the other problems would fall into place.
At this point, Eckert proposed the use of a delay-line storage unit as an alternative to electronic-tube storage. He had calculated that a mercury delay line about
five feet in length would produce a one-millisecond delay; assuming numbers
were represented by electronic pulses of a microsecond duration, a delay line
would be able to store 1,000 such pulses. By connecting the output of the delay
line (using appropriate electronics) to the input, the 1,000 bits of information
would be trapped indefinitely inside the delay line, providing permanent
read/write storage. By comparison, the ENIAC used an electronic "flip-flop"
consisting of two electronic tubes to store each pulse. The delay line would produce a 100-to-1 improvement in the amount of electronics used and make large
amounts of storage feasible.

Eckert's "great new technological invention" provided the agenda for the first
meetings with von Neumann. How could they use the mercury delay line to
overcome the shortcomings of the ENIAC? Very early on, probably shortly after
von Neumann's arrival, the crucial moment occurred when the stored-program
concept was born: that *the computer's storage device would be used to hold both
the instructions of a program and the numbers on which it operated.*

Goldstine has likened the stored-program concept to the invention of the
wheel: It was simple—once one had thought of it. This simple idea would allow
for rapid program set-up by enabling a program to be read into the electronic
memory from punched cards or paper tape in a few seconds; it would be possible to deliver instructions to the control circuits at electronic speeds; it would
provide two orders of magnitude more number storage; and it would reduce
the tube count by 80 percent. But most significantly, it would enable a program
to treat its own instructions as data. Initially this was done to solve a technical
problem associated with handling arrays of numbers, but later it would be used
to enable programs to create other programs—planting the seeds for programming languages and artificial intelligence. But in August 1944, all this lay far in
the future.

Several further meetings with von Neumann took place over the next few
months. The other regular members of the meetings were Eckert, Mauchly,
Goldstine, and Burks—the last named then working on the ENIAC and writing the technical documentation. During these meetings, Eckert and Mauchly's

contributions focused largely on the delay-line research, while von Neumann, Goldstine, and Burks concentrated on the mathematical-logical structure of the machine. Thus there opened a schism in the group between the technologists (Eckert and Mauchly) on the one side and the logicians (von Neumann, Goldstine, and Burks) on the other, which would lead to serious disagreements later on.

Of course, many details needed to be resolved before a complete blueprint of the EDVAC would be available. But at the top level, the architecture—then known as the logical design—was quickly established. Von Neumann played a key role in this aspect of the work, drawing on his training in mathematical logic and his sideline interest in the organization of the brain. During this period, the functional structure of the computer consisting of five functional parts emerged (see diagram, page 83). Von Neumann designated these five units as the central control, the central arithmetic part, the memory, and the input and output organs. The biological metaphor was very strong and, incidentally, is the origin of the term "computer memory"; this term soon displaced "storage," which had been used since Babbage's time.

Another key decision was to use binary to represent numbers. The ENIAC had used decimal numbers, using ten flip-flops to store a single decimal digit. By using binary, the same number of flip-flops could store 10 binary digits—the equivalent of 3 decimal digits. Thus a delay line of 1,024 bits would be able to store 32 "words," each of which could represent an instruction or a number of about 10 decimal digits. Von Neumann estimated that the EDVAC would need between 2,000 and 8,000 words of storage, which would require between 64 and 256 delay lines to implement. This would be a massive amount of equipment, but still far less than the ENIAC.

By the spring of 1945, the plans for EDVAC had evolved sufficiently that von Neumann decided to write them up. His report, entitled *A First Draft of a Report on the EDVAC*, dated 30 June 1945, was the seminal document describing the stored-program computer. It offered the complete logical formulation of the new machine and ultimately was the technological basis for the worldwide computer industry. Although the 101-page report was in draft form, with many references left incomplete, 24 copies were immediately distributed to people closely associated with Project PY. Von Neumann's sole authorship of the report seemed unimportant at the time, but it later led to him being given sole credit for the invention of the modern computer. Today, computer scientists routinely speak of "the von Neumann architecture" in preference to the more prosaic "stored-program concept"; this has done an injustice to von Neumann's co-inventors.

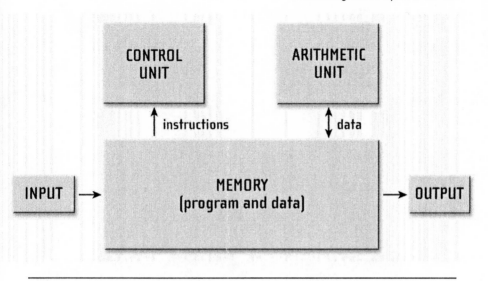

FIGURE 4.1 The stored-program computer of June 1945.

Although von Neumann's *EDVAC Report* was a masterly synthesis, it served to drive further apart the engineers and logicians. For example, in the report von Neumann had pursued the biological metaphor by eliminating all the electronic circuits in favor of logical elements using the "neurons" of brain science. This abstraction made reasoning about a computer far easier (and of course computer engineers now always work at the "gate level"). But Eckert, in particular, believed the *EDVAC Report* skated over the really difficult engineering problems of actually constructing the machine, especially of getting the memory to work. Somehow, Eckert felt, it was all too easy for von Neumann from his lofty height to wish away the engineering problems that would consume him in the years ahead.

Although originally intended for internal circulation to the Project PY group, the *EDVAC Report* rapidly grew famous, and copies found their way into the hands of computer builders around the world. This was to constitute publication in a legal sense, and it eliminated any possibility of getting a patent. For von Neumann and Goldstine, who wished to see the idea move into the public domain as rapidly as possible, this was a good thing; but for Eckert and Mauchly, who saw the computer as an entrepreneurial business opportunity, it was a blow that would eventually cause the group to break up.

The Engineers Versus the Logicians

On 8 May 1945 Germany was defeated and the war in Europe was at an end. On 6 August the first atomic bomb was exploded over Hiroshima. A second bomb was exploded over Nagasaki three days later, and on 14 August Japan surrendered. Ironically, the ENIAC would not be completed for another six weeks, too late to help in the war effort.

Nonetheless, the fall of 1945 was an intensely exciting period, as the ENIAC neared completion. Even the once-indifferent NDRC had begun to take an interest in the machine and requested John von Neumann to prepare reports on both the ENIAC and the EDVAC. Suddenly, computers were in the air. In October the NDRC organized a secret conference at MIT "entirely without publicity and attendance by invitation only" at which von Neumann and his Moore School colleagues disclosed the details of ENIAC and EDVAC to the nascent American computing community. Subsequently von Neumann gave other high-level briefings and seminars on his own. Eckert was particularly irked by von Neumann's unilateral presentations; not only did they have the effect of heaping yet more credit on von Neumann, but they also allowed him to promote his abstract-logical view of computers, which was completely at odds with Eckert's down-to-earth engineering viewpoint.

Eckert clearly felt that von Neumann had taken away some of his own achievement. But if von Neumann's presence overshadowed Eckert's contribution to the invention of the stored-program computer, nothing could take away his unique achievement as chief engineer of the ENIAC. Posterity will surely regard Eckert as the Brunel of the computer age. In particular, the scale of Eckert's engineering achievement clearly sets apart the work of the Moore School from that on the ABC, even if Mauchly did take some ideas from Atanasoff.

Von Neumann, Goldstine, Burks, and some others had a different view. Von Neumann does not seem to have been interested in taking personal credit for the stored-program concept, which he regarded as a team effort, but rather in seeing it disseminated as widely and quickly as possible so that it would come into use for scientific and military applications. Thus, his academic interests came into conflict with Eckert's commercial interests. He noted that the government had paid for the development of the ENIAC and the EDVAC and that therefore it was only fair that these ideas be placed in the public domain. Moreover, as a mathematician rather than an engineer, he believed that it was critically important to distinguish the logical design from the engineering implementation—probably a wise decision given the immaturity of computer engineering and the fact that it was undesirable to lock in too early on an untried technology.

In November 1945, the ENIAC at last sprang into life. It was a spectacular sight. In a basement room at the Moore School, measuring 50 by 30 feet, the units of the ENIAC were arranged in a U-shaped plan around the two longest walls and the shorter end wall. All together there were forty individual units, each two feet wide by two feet deep by eight feet tall. Twenty of the units were the "accumulators," each of which contained 500 tubes and stored a ten-digit decimal number. Three hundred twinkling neon lights on the accumulators made for a science-fiction-like sight when the room was darkened. The other racks around the room included control units, circuits for multiplication and division, and equipment for controlling the IBM card readers and punches so that information could be fed into the machine and the results punched out. Two great twenty-horsepower blowers were installed to draw in cool air and expel the 150 kilowatts of heat that were produced by the machine.

By the spring of 1946, with the ENIAC completed and the war over, there was little to hold the Moore School computer group together. There was mounting tension between Eckert and Mauchly on the one hand, and von Neumann, Goldstine, and Burks on the other. Eckert's resentment over von Neumann's sole authorship of the *EDVAC Report* continued to simmer, there were squabbles about patents, and an inept new Moore School administration all contributed to the breakup of the group.

A new administrative regime at the Moore School took over in early 1946, when the dean, Harold Pender, appointed Irwin Travis as research director. A former assistant professor at the Moore School who had risen to the rank of commander during the war, Travis had spent the war supervising contracts for the Naval Ordnance Laboratory. Travis was a highly effective administrator (he was later to head up Burroughs's computer research), but he had an abrasive personality and soon ran into conflict with Eckert and Mauchly over patents by requiring them to assign all future patent rights to the university. This Eckert and Mauchly absolutely refused to do and instead tendered their resignations.

Eckert could well afford to leave the Moore School because he was not short of offers. First, von Neumann had returned at the end of the war to the Institute for Advanced Study as a full-time faculty member, where he had decided to establish his own computer project. Goldstine and Burks had elected to join him in Princeton, and the offer was extended to Eckert. Even more tempting was an offer from Thomas Watson Sr. at IBM. Watson had caught wind of the computer and, as a defensive research strategy, had decided to establish an electronic computer project. He made an offer to Eckert. However, Eckert was convinced that there was a basis for a profitable business making computers, and so in partnership with Mauchly he secured venture capital. In March 1946 in Philadelphia they set up in business as the Electronic Control Company to build computers.

The Moore School Lectures

Meanwhile, there was intense interest in the ENIAC from the outside world. On 16 February 1946, it had been inaugurated at the Moore School. The event attracted considerable media attention for what was still an abstruse subject. The ENIAC was featured in the movie newsreels around the nation, and it received extensive coverage from the press as an "electronic brain." Apart from its gargantuan size, the feature the media found most newsworthy was its awesome ability to perform 5,000 operations in a single second. This was a thousand times faster than the Harvard Mark I, the only machine with which it could be compared and a machine that had captured the public imagination several years before. The inventors liked to point out that ENIAC could calculate the trajectory of a speeding shell faster than the shell could fly.

Computers were receiving scientific as well as public attention. Scientists were on the move to learn about wartime developments. This was especially true for the British, for whom travel to the United States had been impossible during the war. One of the first British visitors to the Moore School was the distinguished computing expert L. J. Comrie; he was followed by "two others connected with the British Post Office Research Station" (in fact, people from the still top-secret codebreaking operation). Then came visitors from Cambridge University, Manchester University, and the National Physical Laboratory—all soon to embark on computer projects. Of course, there were many more American visitors than British, and there were many requests from university, government, and industrial laboratories for their researchers to spend a period working at the Moore School. In this way they would be able to bring back the technology of computers to their employers. All this attention threatened to overwhelm the Moore School, which was already experiencing difficulties with the exodus of people to join von Neumann in Princeton, and to work at Eckert and Mauchly's company.

Dean Pender, however, recognized the duty of the school to ensure that the knowledge of stored-program computing was effectively transmitted to the outside world. He decided to organize a summer school for thirty to forty invitation-only participants, one or at most two representatives from each of the major research organizations involved with computing. The Moore School

Lectures, as the course later became known, took place over eight weeks from 8 July to 31 August 1946. The list of lecturers on the course read like a *Who's Who* of computing of the day. It included luminaries such as Howard Aiken and von Neumann, who made guest appearances, as well as Eckert, Mauchly, Goldstine, Burks, and several others from the Moore School who gave the bread-and-butter lectures.

The students attending the Moore School Lectures were an eclectic bunch of established and promising young scientists, mathematicians, and engineers. For the first six weeks of the course, the students worked intensively for six days a week in the baking Philadelphia summer heat, without benefit of air-conditioning. There were three hours of lectures in the morning and an afternoon-long seminar after lunch. Although the ENIAC was well explained during the course, for security reasons only limited information was given on the EDVAC's design (because Project PY was still classified). However, security clearance was obtained toward the end of the course, and in a darkened room the students were shown slides of the EDVAC block diagrams. The students took away no materials other than their personal notes. But the design of the stored-program computer was so classically simple that there was no doubt in the minds of any of the participants as to the pattern of computer development in the foreseeable future.

Maurice Wilkes and EDSAC

It is possible to trace the links between the Moore School and virtually all the government, university, and industrial laboratories that established computer projects in America and Britain in the late 1940s. Britain was the only country other than the United States not so devastated by the war that it could establish a serious computer research program. For a very interesting reason, the first two computers to be completed were both in England, at Manchester and Cambridge universities. The reason was that none of the British projects had very much money. This forced them to keep things simple and thus to avoid the engineering setbacks that beset the grander American projects.

Britain had in fact already built up significant experience in computing during the war in connection with codebreaking at the Bletchley Park cryptography center in the English midlands. A mechanical computer—specified by the mathematician Alan Turing—was operational in 1940, and this was followed by an electronic machine, the Colossus, in 1943. Although the codebreaking activity remained completely secret until the mid-1970s, it meant that there were people available who had both an appreciation of the potential for electronic computing and the technical know-how. Turing, for example, established the computing project at the National Physical Laboratory, while Max Newman, one of the prime movers of the Colossus computer, established the computer project at Manchester.

The Manchester computer was the first to become operational. Immediately after the end of the war, Newman became professor of pure mathematics at Manchester University and secured funding to build an EDVAC-type computer. He persuaded a brilliant radar engineer, F. C. (later Sir Frederic) Williams, to move from the Telecommunications Research Establishment (the British radar development laboratories) to take a chair in electrical engineering at Manchester.

Williams decided that the key problem in developing a computer was the memory technology. He had already developed some ideas at the Telecommunications Research Establishment in connection with the Moving Target Indicator problem, and he was familiar with the work at the MIT Radiation Laboratory and the Moore School. With a single assistant to help him, he developed a simple memory system based around a commercially available cathode-ray tube. With guidance from Newman, who "took us by the hand" and explained the principles of the stored-program computer, Williams and his assistant built a tiny computer to test out the new memory system. There were no keyboards or printers attached to this primitive machine—the program had to be entered laboriously, one bit at a time, using a panel of press-buttons, and the results had to be read directly in binary from the face of the cathode-ray tube. Williams later recalled:

> In early trials it was a dance of death leading to no useful result, and what was even worse, without yielding any clue as to what was wrong. But one day it stopped and there, shining brightly in the expected place, was the expected answer. It was a moment to remember . . . and nothing was ever the same again.

The date was Monday, 21 June 1948, and the "Manchester Baby Machine" established incontrovertibly the feasibility of the stored-program computer.

As noted earlier, Comrie had been the first British visitor to the Moore School, in early 1946. Comrie had taken away with him a copy of the *EDVAC Report*, and on his return to England he visited Maurice Wilkes at Cambridge University. Wilkes, a thirty-two-year-old mathematical physicist, had recently returned from war service and was trying to reestablish the computing laboratory at Cambridge. The laboratory had been opened in 1937, but it had been overtaken by the war before it could get into its stride. It was already equipped with a differential analyzer and desk calculating machines, but Wilkes recognized that he also needed to have an automatic computer. Many years later he recalled in his *Memoirs:*

> In the middle of May 1946 I had a visit from L. J. Comrie who was just back from a trip to the United States. He put in my hands a document written by J. von Neumann on behalf of the group at the Moore School and entitled "Draft report on the EDVAC." Comrie, who was spending the night at St John's College, obligingly let me keep it until next morning. Now, I would have been able to take a xerox copy, but there were then no office copiers in existence and so I sat up late into the night reading the report. In it, clearly laid out, were the principles on which the development of the modern digital computer was to be based: the stored program with the same store for numbers and instructions, the serial execution of instructions, and the use of binary switching circuits for computation and control. I recognized this at once as the real thing, and from that time on never had any doubt as to the way computer development would go.

A week or two later, while Wilkes was still pondering the significance of the *EDVAC Report*, he received a telegram from Dean Pender inviting him to attend the Moore School Lectures during July and August. There was no time to organize funds for the trip, but Wilkes decided to go out on a limb, pay his own way, and hope to get reimbursed later. In the immediate postwar period transatlantic shipping was tightly controlled. Wilkes made his application for a berth but had heard nothing by the time the course had started. Just as he was "beginning to despair" he got a phone call from a Whitehall bureaucrat and was offered a passage at the beginning of August.

After numerous hitches Wilkes finally disembarked in the United States and enrolled in the Moore School course on Monday, 18 August, with the course having just two weeks left to run. Wilkes, never short on confidence, decided that he was "not going to lose very much in consequence of having arrived late," since much of the first part of the course had been given over to material on numerical mathematics with which he was already familiar. As for the EDVAC, Wilkes found the "basic principles of the stored program computer were easily grasped." Indeed, there was nothing more to grasp than the basic principles: All the details of the physical implementation were omitted, not least because the design of the EDVAC had progressed little beyond paper design studies. Even the basic memory technology had not yet been established: True, delay lines and cathode-ray tubes had been used for storing radar signals for a few milliseconds, but this was quite different from the bit-perfect storage of minutes' or hours' duration needed by digital computers.

After the end of the Moore School course, in the few days that were left to him before his departure for Britain, Wilkes visited Harvard University in Cambridge, Massachusetts. There he was received by Howard Aiken and saw both the Harvard Mark I grinding out tables and a new relay machine, the Mark II, under construction. To Wilkes, Aiken now appeared to be one of the older generation, hopelessly trying to extend the technological life of relays while resisting the move to electronics. Two miles away, at MIT, he saw the new electronic differential analyzer, an enormous machine of about 2,000 tubes, thousands of relays, and 150 motors. It was another dinosaur. Wilkes was left in no doubt that the way forward had to be the stored-program computer.

Sailing back on the *Queen Mary*, Wilkes began to sketch out the design of a "computer of modest dimensions very much along the lines of the EDVAC Proposal." Upon his return to Cambridge in October 1946, he was able to direct the laboratory's modest research funds toward building a computer without the necessity for formal fundraising. From the outset, he decided that he was interested in *having* a computer, rather than in trying to advance computer engineering technology. He wanted the laboratory staff to become experts in using computers—in programming and mathematical applications—rather than in building them. In line with this highly focused approach, he decided to use delay-line storage rather than the much more difficult, though faster, CRT-based storage. This decision enabled Wilkes to call the machine the Electronic Delay Storage Automatic Calculator; the acronym EDSAC was a conscious echo of its design inspiration, the EDVAC.

During the war Wilkes had been heavily involved in radar development and pulse electronics, so the EDSAC presented him with few technical problems. Only the mercury delay-line memory was a real challenge, but he consulted a colleague who had successfully built such a device for the admiralty during the war and who was able to specify with some precision a working design. Wilkes copied these directions exactly. This decision was very much in line with his general philosophy of quickly getting a machine up and running on which he could "try out real programs instead of dreaming them up for an imaginary machine."

By February 1947, working with a couple of assistants, Wilkes had built a successful delay line and stored a bit pattern for long periods. Buoyed up by the confidence of a mercury delay-line memory successfully implemented, he now pressed on with the construction of the full machine. There were many logisti-

cal as well as design problems; for example, the supply of electronic components in postwar Britain was very uncertain and required careful planning and
advanced ordering. But luck was occasionally on his side, too:

> I was rung up one day by a genial benefactor in the Ministry of Supply who
> said that he had been charged with the disposal of surplus stocks, and what
> could he do for me. The result was that with the exception of a few special
> types we received a free gift of enough tubes to last the life of the machine.

Wilkes's benefactor was amazed at the number of electronic tubes—several
thousand—that Wilkes could make use of.

Gradually the EDSAC began to take shape. Eventually 32 memory delay lines
were included, housed in two thermostatically controlled ovens to ensure temperature stability. The control units and arithmetic units were held in three
long racks standing six feet tall, all the tubes being open to the air for maximum
cooling effect. Input and output were by means of British Post Office teletype
equipment. Programs were punched on telegraph tape, and results were printed
on a teletype page printer. All together the machine contained 3,000 tubes and
consumed 30 kilowatts of electric power. It was a very large machine, but it had
only one-sixth the tube count of the ENIAC and was correspondingly smaller.

In early 1949 the individual units of the EDSAC were commissioned and the
system, equipped with just four temperamental mercury delay lines, was assembled. Wilkes and one of his students wrote a simple program to print a table of
the squares of the integers. On 6 May 1949, a thin ribbon of paper containing
the program was loaded into the computer; half a minute later the teleprinter
sprang to life and began to print 1, 4, 9, 16, 25 . . . The world's first practical
stored-program computer had come to life, and with it the dawn of the computer age.

5

:: THE COMPUTER BECOMES A BUSINESS MACHINE

AN APOCRYPHAL tale about Thomas J. Watson Sr., the onetime president of IBM, claims that, probably around 1949, he decided that there was a market for no more than about a dozen computers and that IBM had no place in that business. This anecdote is usually told to show that the seemingly invincible leader of IBM was just as capable of making a foolish decision as the rest of the human race. Conversely, the story portrays as heroes those few individuals, such as Eckert and Mauchly, who foresaw a big market for computers. There is perhaps some justice in this verdict of posterity on the conservative Watson versus the radical Eckert and Mauchly. Another way of interpreting these responses to the computer is as the rational versus the irrational, based on business acumen and past experience. Thus Watson was rational but wrong, while Eckert and Mauchly were irrational but right. It could just as easily have been the other way around—and then Watson would have still been a major business leader, but not many people would have heard of Eckert and Mauchly.

What really happened in the 1950s, as suggested by the title of this chapter, is that the computer was reconstructed—mainly by computer manufacturers and business users—to be an electronic data-processing machine rather than a mathematical instrument. Once IBM had recognized this change, in about 1951, it altered its sales forecasts, rapidly reoriented its R&D, manufacturing, and sales organizations, and used its traditional business strengths to dominate the industry within a period of five years. This was a heroic achievement too, though not one of which myths are made.

In the late 1940s and early 1950s, some thirty firms entered the computer business in the United States. The only other country to develop a significant computer industry at this time was the United Kingdom, where, in an effort to capture a share of what was seen as a promising new postwar industry, about ten computer companies came into being—size for size, a comparable performance with the United States. Britain even had the world's first computer on the market. This was a machine known as the Ferranti Mark I, based on the Manchester University computer and delivered in February 1951. Unfortunately the enthusiasm for manufacturing computers in Britain was not matched with an enthusiasm for using them by its old-fashioned businesses, and by the end of the decade Britain's computer industry was fighting for survival.

As for the rest of the world, all the industrially developed countries of continental Europe—Germany, France, Holland, and Italy—were so ravaged by World War II that they would not be capable of entering the computer race until the 1960s. The Eastern bloc and Japan were even slower to develop computers. Hence, for the 1950s, U.S. firms had the biggest market for computers—the United States—to themselves, and from this springboard they were able to dominate the world.

There were actually three types of firms that entered the computer industry: electronics and control equipment manufacturers, office-machine companies, and entrepreneurial startups. The electronics and control equipment manufacturers—including RCA, General Electric, Raytheon, Philco, Honeywell, Bendix, and several others—were the firms for whom the computer was the most "natural" product. They were well used to selling high-cost electronic capital goods, such as radio transmitters, radar installations, X-ray equipment, and electron microscopes. The computer, for which these firms coined the term "mainframe," was simply another high-value item in their product line.

The second class of firm entering the computer market was the business-machine manufacturers such as IBM, Remington Rand, Burroughs, NCR, Underwood, Monroe, and Royal. They lagged the electronics firms by a couple of years because, quite rationally, they did not see the manufacture of mathematical computing instruments as being an appropriate business for them to enter. It was not until Eckert and Mauchly produced the UNIVAC computer in 1951, proving that computers had a business application, that they entered the race with conviction.

The third class of firm to enter the computer business—and in many ways the most interesting—was the entrepreneurial start-up. The period from 1948 to 1953 was a narrow window of opportunity when it was possible to join the computer business at a comparatively low cost, certainly less than a million dollars. A decade later, because of the costs of supplying and supporting peripherals and software, one would need $100 million to be able to compete with the established mainframe computer companies. The first entrepreneurial computer firms were Eckert and Mauchly's Electronic Control Company (set up in 1946) and Engineering Research Associates (ERA, also set up in 1946). These firms were soon taken over by Remington Rand to become its computer division: Eckert and Mauchly became famous as the pioneers of a new industry, while William Norris, one of the founders of ERA, moved on to found another company—the highly successful Control Data Corporation.

Following quickly on the heels of the Eckert-Mauchly company and ERA came a rush of computer start-ups, sometimes funded with private money, but more often the subsidiaries of established electronics and control companies. Firms emerging in the early fifties included ElectroData, the Computer Research Company, Librascope, and the Laboratory for Electronics. The luckiest of these firms were taken over by office-machine companies, but most simply faded away, unable to survive in the intensely competitive environment. The story of the Eckert-Mauchly company illustrates well the precarious existence shared by virtually all the start-up companies.

"More Than Optimistic": UNIVAC and BINAC

When Eckert and Mauchly established their Electronic Control Company in March 1946, they were almost unique in seeing the potential for computers in business data processing, as opposed to science and engineering calculations. Indeed, even before the end of the war, Eckert and Mauchly had paid several visits to the Bureau of the Census in Washington, canvassing the idea of a computer to help with census data processing, though nothing came of the idea at the time. The vision of computerized data processing probably came more from Mauchly than from Eckert. Thus, at the Moore School Lectures in the summer of 1946, while Eckert had talked about the technical details of computer hardware, Mauchly was already giving a lecture on sorting and collating—topics that really belonged in the world of punched-card business machines, not mathematical computation.

When Eckert and Mauchly set out to build their business, there were few venture-capital firms of the kind we are now familiar with. They therefore decided that the most effective way to fund the firm would be to obtain a firm order for their computer and use a drip feed of incremental payments to develop the machine. Those wartime visits to the Census Bureau paid off. In the spring of 1946, the Bureau of the Census agreed to buy a computer for a total purchase price of $300,000. Eckert and Mauchly, who were hopelessly optimistic about the cost of developing their computer, thought it would cost $400,000 to develop the machine (much the same as ENIAC), but $300,000 was the best deal they could negotiate. They hoped to recover their losses in subsequent sales.

In fact, the computer would cost closer to a million dollars to develop. As one historian has noted, "Eckert and Mauchly were more than optimistic, they were naive." Perhaps it was a foolish way to start a business, but, like entrepreneurs before and after, Eckert and Mauchly wished away the financial problems in order to get on with their mission. If they had stopped to get financially rewarding terms, the project would likely have been strangled at birth. The contract to build an EDVAC-type machine for the Census Bureau was signed in October 1946, and the following spring the computer was officially named UNIVAC, for UNIVersal Automatic Computer.

Eckert and Mauchly started operations in downtown Philadelphia, taking the upper two stories of a men's clothing store at 1215 Walnut Street. A narrow-fronted building, though running deep, it was more than sufficient for the half-dozen colleagues they had persuaded to join them from the Moore School. By the autumn of 1947, there were about twelve engineers working for the company, all engaged in the various parts of the UNIVAC.

Eckert and Mauchly had embarked upon a project whose enormousness only began to reveal itself during this first year of the company's operation. Building the UNIVAC was much more difficult than simply putting together a number of ready-made subsystems, as would be the case when building a computer a decade later. Not even the basic memory technology had been established, so that one group of engineers was working on a mercury delay-line memory—and in the process of discovering that it was one thing to develop a system for laboratory use, but quite another to produce a reliable system for fuss-free use in a commercial environment.

The most ambitious feature of the UNIVAC was the use of magnetic-tape storage to replace the millions of punched cards used by the Census Bureau and other organizations and businesses. This was a revolutionary idea, which spawned several other projects. In 1947 the only commercially available tape decks were analog ones used in sound recording studios. A special digital magnetic-tape drive had to be developed so that computer data could be put onto the tape, and high-speed servomotors had to be incorporated so that the tape could be started and stopped very rapidly. Then it turned out that ordinary commercial magnetic tape, based on a plastic substrate, stretched unacceptably.

They now had to set up a small chemistry laboratory to develop tape-coating materials, and use a metallic tape that did not stretch. Next a machine had to be developed so that data could be keyed onto the tape, and then a printer so that the contents of a tape could be listed. Later, machines would be needed to copy decks of cards onto magnetic tape, and vice versa. In many ways the magnetic-tape equipment for the UNIVAC was both the most ambitious and the least successful aspect of the project.

Eckert and Mauchly were now both working intensely hard for very long hours, as was every engineer in the company. The company was also beset by financial problems, and to keep the operation afloat Eckert and Mauchly needed more orders for UNIVACs and more advance payments. These orders were not easy to come by. However, a few months earlier, Mauchly had been hired by the Northrop Aircraft Corporation to consult on guidance computers for the Snark long-range guided missile that the company was developing. Northrop decided it needed a small airborne computer, and Eckert and Mauchly were invited to tender for the development contract for this machine—to be called the BINAC (for binary automatic computer).

The BINAC would be a very different order of machine than the UNIVAC. It would be much smaller; it would be a scientific machine, not a data processor; and it would not use magnetic tape—perhaps the biggest task in Eckert and Mauchly's portfolio. The contract to develop the BINAC would keep the company afloat, and yet it would also detract greatly from the momentum of the UNIVAC. But there was no real choice: It was the BINAC or go out of business. In October 1947 they signed a contract to develop the BINAC for $100,000, of which $80,000 was paid in advance.

Despite the distraction of the BINAC, the UNIVAC remained central to the long-term ambition of the company, and the search for UNIVAC orders continued. The most promising prospect was the Prudential Insurance Company, which Mauchly had first approached in 1946 to no avail, but whose interest had now started to revive. The Prudential was a pioneer in the use of office machinery and had even developed its own punched-card system in the 1890s, long before IBM punched-card machines were commercially available; so it was no surprise that it would become an early adopter of the new computer technology.

The Prudential's computer expert was a man named Edmund C. Berkeley, who was to write the first semipopular book on computers, *Giant Brains*, published in 1949. Early in 1947, Mauchly and Berkeley had tried to convince the Prudential board to order a UNIVAC, but the company was unwilling to risk several hundred thousand dollars to buy an unproved machine from an unproved company. However, after several months of negotiation, the Prudential agreed to pay $20,000 for Mauchly's consulting services, with an option to buy a machine when it was developed. It wasn't much, but every little bit helped, and there was after all the distant prospect that the Prudential would eventually buy a machine.

By the end of 1947 the Eckert and Mauchly operation was spending money much faster than it was recovering it, and it needed further investment capital. In order to attract investors they incorporated the company as the Eckert-Mauchly Computer Corporation (EMCC). Eckert and Mauchly were desperately seeking two things, investors and contracts for the UNIVAC.

In the spring of 1948, the A. C. Nielson market research firm in Chicago agreed to buy the second UNIVAC for $150,000, and toward the end of the year the Prudential offered to buy the third for the same price. This was a ludicrously unrealistic price tag, although, as always, Eckert and Mauchly wished away the problems. Buoyed up by the prospect of firm orders, more engineers, technicians, and draftsmen were recruited, and the company moved to larger premises, taking the seventh and eighth floors of the Yellow Cab Building on the edge of central Philadelphia. Eckert and Mauchly were now employing forty people, including twenty engineers, and they needed an additional $500,000 in working capital. This financial support would shortly be offered by the American Totalisator Company of Baltimore.

It so happened that Eckert and Mauchly's patent attorney was friendly with Henry Strauss, a vice president of American Totalisator. Strauss was a remarkable inventor. In the 1920s, he had perfected and patented a totalisator machine for racetrack betting that soon dominated the market. The totalisator was, in its way, one of the most sophisticated mechanical computing systems of its era. Strauss immediately saw that the UNIVAC represented a technological breakthrough that might prove a source of future competition in the totalisator field. In August 1948 he persuaded his company to take a 40 percent shareholding in the Eckert-Mauchly Computer Corporation. Strauss, an inventor himself, had the wisdom to leave Eckert and Mauchly in full control of their company, while he would provide mature guidance by serving as chairman of the board. American Totalisator offered about $500,000 for its share of EMCC and offered further financing through loans. It was a truly generous offer, and Eckert and Mauchly grasped it thankfully.

With financial stability at last, EMCC had turned the corner. True, virtually every deadline and schedule had been missed and the future promised nothing different, but this was to some extent to be expected in a high-tech company. The important thing was that the money from American Totalisator and the contracts with the Census Bureau, Northrop, Nielson, and Prudential ensured that the company could expand confidently. The following spring, it moved for the third time in as many years, this time to a self-contained two-story building in North Philadelphia. A former knitting factory, when emptied of its plant, the interior was so vast that it was "difficult . . . to believe that we would ever outgrow this facility." Situated at the foot of a hill, the area was so hot in the summer that they called it Death Valley.

As soon as they moved in, there was a burst of activity to complete the BINAC. The first American stored-program computer to operate, the BINAC was never a reliable machine. In the early summer a visiting Northrop engineer reported back that it was operating about one hour a week, "but very poorly." It was only operating a little more reliably when it was finally shipped off to Northrop in September 1949. It never really performed to expectation, but then few early computers did, and Northrop was happy enough to pay the outstanding $20,000 of the $100,000 purchase price.

With the BINAC out of the way, Eckert and Mauchly could at last focus single-mindedly on the UNIVAC, for which three more sales had now been secured. EMCC was now a bustling company in its new headquarters, with 134 employees and contracts for six UNIVAC systems totaling $1.2 million. Things had never looked brighter. A few weeks later, however, Eckert and Mauchly received the news that Henry Strauss had been killed when his privately owned twin-engine airplane crashed. Since Strauss was the prime mover behind American Totalisator's investment in EMCC, when he died so did American Totalisator's commitment to the company. It pulled out and required its loans to be repaid. EMCC was immediately thrown into financial disarray, and all the old uncertainties resurfaced. It was as though someone had turned out the light.

IBM: Evolution, Not Revolution

Meanwhile IBM, the company that was to become Eckert and Mauchly's prime competitor, was beginning to respond to the computer. As mentioned at the chapter's opening, there is a myth that IBM made a fateful decision shortly after the war not to enter the computer business, made a swift reversal of policy about 1950, and by the mid-1950s had secured a position of dominance. What is most beguiling about this myth is that it is sufficiently close to the truth that people take it at face value. Yet where it is misleading is that it portrays IBM as sleeping soundly at the start of the computer race. The truth is more complex.

IBM had several electronics and computer development projects in the laboratories in the late 1940s; what delayed them being turned into products was the uncertainty of the market.

Immediately after the war, IBM (like the other accounting-machine companies such as Remington Rand, Burroughs, and NCR) was trying to respond to three critical business challenges: product obsolescence, electronics, and the computer. The first and most critical challenge was product obsolescence. All IBM's traditional electromechanical products were in danger of becoming obsolete because during the war most of its R&D facilities had been given over to military contracts such as the development of gun aimers and bomb sights. For the very survival of IBM's business, revamping its existing products had to be the top postwar priority.

The second challenge facing the office-machine companies was the threats and opportunities presented by electronics technology. Electronics had been greatly accelerated by the war, and IBM had itself gained some expertise by constructing code-breaking machines and radio equipment. The race was now on to incorporate electronics into the existing products. The third challenge was the stored-program computer. But in 1946, when the computer was a specialized mathematical instrument of no obvious commercial importance, it had to be the lowest priority; anything else would have been irrational. IBM institutionalized its attitude to electronics and computers through the slogan "evolution not revolution." By this, it meant that it would incorporate electronics into existing products to make them faster, but they would not otherwise be any different. Other industry pundits likened the office-machine manufacturers' attitude to electronics as similar to that of aircraft manufacturers toward the jet engine: The new technology would make their products faster but would not change their function.

Thomas Watson Sr., often portrayed as an electromechanical diehard, was in fact among the first to see the future potential of electronics. As early as October 1943 he had instructed his R&D chief to "find the most outstanding Professor on electronics and get him for IBM." Although cost was no object, no suitable candidate could be found because all the electronics experts were committed to the war effort at that time. Nothing more might have happened for a few years had there not been that unpleasant incident in August 1944 when the Mark I calculator was inaugurated at Harvard University, and Howard Aiken refused to acknowledge IBM's role in inventing and building the machine.

Watson decided to slight Aiken's Mark I by building a more powerful, electronic calculator. In March 1945 IBM hired Professor Wallace Eckert (no relation to Presper) to establish the Watson Computer Laboratory at Columbia University and develop a "supercalculator" to be known as the Selective Sequence Electronic Calculator (SSEC). Watson's objective—apart from thumbing his nose at Aiken—was to ensure that IBM had in the Watson Computation Laboratory a test bed for new ideas and devices. It was not expected that the SSEC—or any other computing machine—would develop into a salable product. The decision to build the SSEC was made a couple of months before the publication of von Neumann's *EDVAC Report.* For this reason, the SSEC was a one-of-a-kind development that never became part of the mainstream of stored-program computing.

Meanwhile, IBM's product development engineers were responding to the less glamorous challenge of incorporating electronics into the company's existing products. The first machine to receive this treatment was the Model 601 multiplying punch, first marketed in 1934. The problem with this machine was that its electromechanical multiplier could handle only about 10 numbers per minute. The heart of the machine was ripped out and replaced with electronics—giving the machine a tenfold speed improvement, and it was now able to perform about 100 multiplications per minute. Containing approximately 300 vacuum tubes, the machine was first demonstrated at the National Business Show in New York in September 1946. It was marketed as the IBM Model 603, and about 100 machines were produced before it was replaced two years later by the much more versatile Model 604 calculating punch. The latter electronic workhorse went on to sell a staggering 5,600 units over a ten-year period. With 1,400 tubes and a limited programming capability, the 604 provided "speed and flexibility of operation unmatched by any calculator in the market for some time." More than that, the 604 also provided the heart of a powerful computing setup known as the CPC—the Card Programmed Calculator.

The CPC was developed in 1947 in a cooperative effort between IBM and one of its West Coast customers, Northrop Aircraft, the same company that had commissioned the BINAC from Eckert and Mauchly. Northrop was a heavy user of IBM equipment for calculating missile trajectories and took early delivery of a Model 603 multiplier. In order to improve the flexibility of the 603, the machine was hooked up to a specially developed storage unit and other punched-card equipment. The result was a highly effective calculating setup capable of about one thousand operations per second. Very quickly, news of Northrop's installation began to spread, particularly among West Coast aerospace companies, and by the end of 1948 IBM had received a dozen requests for a similar setup. This was the first demonstration that IBM had received that there was a significant market for scientific computing equipment.

IBM product engineers set to work to develop the CPC, in which the more powerful 604 calculating punch replaced the 603 multiplier. Customer deliveries began in late 1949. The CPC was not a true stored-program computer, but it was an effective and very reliable calculating tool in 1949 and 1950, when there were no commercially available computers on the market. Moreover, even when computers did become commercially available, the low cost and superior reliability of the CPC continued to make it the most cost-effective computing system available. Some 700 systems were delivered in the first half of the 1950s, more than the world total of stored-program computers at that date.

Commentators have often failed to realize the importance of the CPC, not least because it was called a "calculator" instead of a "computer." Watson insisted on this terminology because he was concerned that the latter term, which had always referred to a human being, would raise the specter of technological unemployment. It was why he had also insisted on calling the Harvard Mark I and the SSEC calculators. But the fact remained that so far as effective scientific-computing products went, IBM had the lead from the very beginning.

Thus by 1949 IBM had built up an excellent R&D capability in computers. As well as the CPC, there was the SSEC, which was completed in early 1948. Although this was a one-of-a-kind development and not technically a stored-program computer, it was the most advanced and powerful machine available when it was completed. Watson had the machine placed on public view on the ground floor of the IBM headquarters in Manhattan. One overseas visitor wrote:

> This machine was completed and put into operation by IBM at their Headquarters in New York early in 1948. It could be seen from the street by passing pedestrians who affectionately christened it "Poppa." It was a very large machine and contained 23,000 relays and 13,000 [tubes]. . . . The machine in operation must have been the most spectacular in the world. Thousands of neon lamps flashed on and off, relays and switches buzzed away and the tape readers and punches worked continuously.

IBM courted publicity about the machine. Staff writers for *The New Yorker* reported that "[p]eople who just want to look at the calculator, as we did, are always welcome," and were themselves given a conducted tour by Robert Seeber Jr., a senior engineer. The magazine produced a wonderful vignette of the machine that reflected exactly the popular fascination with "electronic brains":

> The principal cerebral parts of the machine are tubes and wires behind glass panels, covering three walls of the room. Two hoppers, looking rather like

oversized mailboxes, stand near the middle of the room. One is the "in" hopper, into which questions are inserted on punched cards or tapes; the other is the "out" hopper, from which, if all goes well, the answer emerges. High on one wall of the room is a large sign reading "THINK," but this admonition isn't addressed to the calculator. . . . The calculator has tackled any number of commercial problems. When we arrived, it was just warming up for one that involved getting more oil out of oil fields. . . . As we moved along, relays began to rattle in a carefree manner and patterns of light danced across the panels. "The oil problem," Seeber said.

The SSEC did a great deal to enhance IBM's reputation as a leader in computer technology. There was, however, no prospect that the SSEC could become a product. It cost far too much ($950,000) and was technically obsolete as soon as it was built. Its real importance to IBM, apart from publicity and image, was to create a cadre of experienced computer engineers.

IBM also had two further full-scale computers under development in 1949: the Magnetic Drum Calculator (MDC) and the Tape Processing Machine (TPM). The MDC was to sell for perhaps a tenth of the price of a UNIVAC, but would give true stored-program computing for the kinds of firms currently using the CPC, as well as for ordinary business users currently using punched-card machines. The key money-saving technology behind it was to use a magnetic drum for the main memory, instead of mercury delay lines or electrostatic storage tubes. The drum memory consisted of a rapidly rotating magnetized cylinder, which made for a very economical and reliable form of memory, but also one that was quite slow.

IBM's Tape Processing Machine was much more on a par with the UNIVAC. As early as 1947, IBM had begun to explore the phenomenon of magnetic-tape recording and its potential as a replacement for punched cards in data processing, but the project moved slowly because of the competing demands of the SSEC. However, in mid-1949 reports of activities at the Eckert-Mauchly operation began to filter through to IBM—first, the successful completion of the BINAC, and second, the news that several orders had been taken for UNIVACs. The latter news convinced IBM to accelerate the TPM so that it would have a large, tape-based data-processing computer that would be competitive with UNIVAC.

If IBM had pressed on with either or both of these computers in 1949 or 1950, it would have dominated the market for data-processing computers much earlier than it did. In fact, these machines did not reach the market for another five years. The delay of the MDC was due largely to conservatism in IBM's marketing. In 1949 the view in IBM's "Future Demands" department was that the MDC was too expensive to compete with ordinary punched-card machines but too slow to compete with the UNIVAC. The delay with the Tape Processing Machine, however, had quite different causes: the start of the Korean War and a strategic decision made by Watson's son, Thomas J. Watson Jr.

By 1950 Watson senior was in his mid-seventies, and while he remained president and chief executive of IBM until 1956, his thirty-five-year-old elder son was being groomed to lead the company. Thomas J. Watson Jr. joined IBM as a young man in 1937, learning the ropes as a salesman. After war service he returned to IBM in 1945, and four years later was made an executive vice president. When the Korean War broke out in June 1950, as in previous emergencies, Watson senior telegraphed the White House and put IBM at the president's disposal. While Watson senior's motive was "primarily patriotic," Watson junior "seized on the idea as an opportunity to speed the company's development of large-scale computers." Market reports indicated that there would be a potential market for at least six large-scale scientific computers with defense contractors, so Watson junior committed the company to developing a computer based on the machine that von Neumann was building at the Institute of Advanced Study.

This would be a very powerful machine indeed; initial estimates put the rental at $8,000 per month—equivalent to a purchase price of $500,000—an estimate that turned out to be just about half of the eventual cost. In order to develop the Defense Calculator, as the machine was called, the resources devoted to the Tape Processing Machine had to be taken away. In effect, Watson's decision to build the Defense Calculator in the summer of 1950, while accelerating IBM's entry into scientific computers, dashed its chances of an early entry into the data-processing market. It was a business error, one that allowed Eckert and Mauchly to capture the early market for data-processing computers with the UNIVAC.

UNIVAC Comes to Life

It will be recalled that in late 1949, Eckert and Mauchly had just learned of the withdrawal of American Totalisator and were urgently seeking a new backer to sustain the momentum of the UNIVAC development and fund the payroll for their 134 employees in Philadelphia. Eckert and Mauchly were, of course, already well known to corporations with an interest in computers. Indeed, when the BINAC had been inaugurated in August 1949, representatives from several companies, including IBM and Remington Rand, had attended.

Early in 1950 Eckert and Mauchly decided that they would have to try to sell the company to IBM. They secured an interview—for an unexplained purpose—with the Watsons, father and son, in their New York office. Tom Watson Jr. recalled the interview:

> I was curious about Mauchly, whom I'd never met. He turned out to be a lanky character who dressed sloppily and liked to flout convention. Eckert, by contrast, was very neat. When they came in, Mauchly slumped down on the couch and put his feet up on the coffee table—damned if he was going to show any respect for my father. Eckert started describing what they'd accomplished. But Dad had already guessed the reason for their visit, and our lawyers had told him that buying their company was out of the question. UNIVAC was one of the few competitors we had, and antitrust law said we couldn't take them over. So Dad told Eckert, "I shouldn't allow you to go on too far. We cannot make any kind of arrangement with you, and it would be unfair to let you think we could. Legally we've been told we can't do it."

Eckert took the point immediately and thanked the Watsons for their time, even though the interview had been unproductive, while "Mauchly never said a word; he slouched out the door after an erect Eckert." Perhaps Watson's recollection says as much about his own values and those of IBM as it does about those of Mauchly. But it was quite plain that Mauchly would never have fit in with the IBM culture. In any case the Watsons must surely have judged that Eckert and Mauchly would bring very little to IBM that it did not already have under development; by 1950 the company had well over a hundred R&D workers engaged in electronics and computer research, an effort that comfortably matched that of Eckert and Mauchly's company.

Remington Rand and NCR were also interested in making some kind of agreement with Eckert and Mauchly, and the entrepreneurs were "so desperate that they were ready to consider the first reasonable offer." Remington Rand got its bid in first.

In fact, even before the end of World War II, James Rand Jr., the charismatic president of Remington Rand, had drawn up a grand postwar expansion plan. He had been deeply impressed by the application of science to wartime activities, and he intended that after the war Remington Rand would develop a whole raft of high-technology products, many of them incorporating electronics. These would include microfilm recorders, xerographic copiers, industrial television systems, and so on. In late 1947 he recruited General Leslie R. Groves, the famous administrative chief of the Manhattan Project, to coordinate the research effort and head Remington Rand's development laboratories. Groves's involvement with atomic weapons research during the war had made him aware of the ENIAC and of Eckert and Mauchly's company; he well understood that computers would come to play an important role in the office.

James Rand invited Eckert and Mauchly down to his retreat in Florida, where they were royally entertained aboard his yacht, and he made them an offer: Remington Rand would repay all of American Totalisator's investment in the company—amounting to $438,000—would pay Eckert, Mauchly, and their employees $100,000 for the stock they owned, and would retain Eckert and Mauchly at a salary of $18,000 per year. It was not a wonderful offer, for they would now become mere employees of the company they had founded, but there was no real alternative. Eckert and Mauchly accepted, and the Eckert-Mauchly Computer Corporation became a wholly owned subsidiary of Remington Rand. Although nominally under the direction of Groves, the general wisely avoided disrupting the UNIVAC operation or relocating the engineering program. Rather, he let things roll along with the considerable momentum Eckert and Mauchly had already built up.

While Remington Rand took a relaxed view of the firm's engineering development, it rapidly set the chronic financial condition to rights. After one of their visits to the Eckert-Mauchly operation, Rand and Groves and their financial advisers realized that the UNIVAC was being sold at perhaps a third of its true cost. First they approached the Census Bureau to try to renegotiate the price, threatening somewhat abrasively to cancel the order otherwise. The Census Bureau threatened to countersue, so the original price of $300,000 had to stand. The Prudential and Nielson machines had been offered at the giveaway price of $150,000, whereas Remington Rand now realized it could not make a profit on anything less than $500,000. In order to extricate itself from their contractual obligations, Remington Rand threatened to involve the Prudential and Nielson in court proceedings that would hold up delivery of the UNIVAC for several years, by which time it would be obsolete. Nielson and Prudential canceled their contracts and had their money returned. Eventually both bought their first computers from IBM.

While these negotiations were going on in the background, Eckert and Mauchly focused on completing the first UNIVAC. By the spring of 1950 it was at last beginning to take shape: "first, one bay was set up on the floor, then another, and another." As ever, Eckert was the hub around which the whole project revolved. Still a young man, in his early thirties, he worked longer hours than anyone:

Each day he worked a little later and consequently was forced to arrive a little later in the morning. As the days went by, his work period slipped later and later into the day and then the evening. Finally he had to take a day off so that he could get re-phrased and come in with everyone else in the morning.

Eckert had always had a curious habit of having to work out his ideas by talking them through with someone who could act as a sounding board. In the ENIAC days it had been Mauchly who had been the sounding board, but now it was usually one of the other engineers:

Pres's nervous energy was so great he couldn't sit in a chair or stand still while he was thinking. He usually crouched on top of a desk or else paced back and forth. In many of his discussions with [one of his engineers], the pair started out at the computer test site, walked to the back of the second floor, then down a flight of steps to the first floor. After an hour or so, they would arrive back at the test site, having made a circular tour of the building, but completely oblivious to the fact that they had done so. The intensity of their discussions locked out all distractions.

By contrast, Mauchly had a laid-back, laconic style: "John was always a great morale booster for everyone. His sense of humor penetrated the black clouds. He spoke in a slow, halting way as he related one interesting anecdote after another."

By the summer of 1950 the UNIVAC subsystems were close to completion and were being tested in the baking Philadelphia heat. The equipment was located on the second floor of the UNIVAC building, beneath an uninsulated flat black roof. There was no air-conditioning, and even if there had been it could not have begun to dissipate the 120 kilowatts of heat produced by the 5,000-tube machine. First neckties and jackets were discarded; a few days later further layers of clothing were dispensed with; and finally "shorts and undershirts became the uniform of the day." One engineer recalled of another: "He had a half dozen soda bottles sitting on the edge of his desk and every once in a while, he reached for a soda bottle, placed it over the top of his head, and poured. The bottles were filled with water!"

By the beginning of 1951, UNIVAC was functioning as a computer and would soon be ready to commence its acceptance tests for the Census Bureau. Before this was possible, however, some basic software had to be developed. The programming team was initially led by Mauchly, but was later run by a programmer recruited from the Harvard Computation Laboratory, Grace Murray

Hopper, who would become the driving force behind advanced programming techniques for commercial computers and the world's foremost female computer professional.

On 30 March 1951 UNIVAC ran its acceptance tests, performing all together some seventeen hours of rigorous computing without a fault. The machine then became the property of the Census Bureau. Over the following year, two more UNIVACs were completed and delivered to the government, and orders for a further three were taken.

In late 1952, Remington Rand achieved a spectacular publicity stunt for UNIVAC. The company persuaded the CBS television network to use UNIVAC to predict the outcome of the presidential election. Some months before the election, John Mauchly, with the help of a statistician from the University of Pennsylvania, devised a program that would use the early returns from a number of key states to predict the result, based on the corresponding voting patterns in 1944 and 1948.

On election night CBS cameras were installed in the UNIVAC building in Philadelphia with reporter Charles Collingwood as the man on the spot. At CBS headquarters, Walter Cronkite served as anchorman, while in the studio a dummy UNIVAC console was installed for dramatic effect—the drama being heightened by flickering console lights, driven by nothing more sophisticated than blinking Christmas tree lights.

The UNIVAC printed its first prediction at 8:30 P.M.:

IT'S AWFULLY EARLY, BUT I'LL GO OUT ON A LIMB
UNIVAC PREDICTS—WITH 3,398,745 VOTES IN—

	STEVENSON	EISENHOWER
STATES	5	43
ELECTORAL	93	438
POPULAR	18,986,436	32,915,049

THE CHANCES ARE NOW 00 TO 1 IN FAVOR OF THE ELECTION OF EISENHOWER.

UNIVAC was forecasting a landslide for Eisenhower—in complete contrast to the Gallup and Roper opinion polls taken the previous day, which had predicted a close race. A member of the UNIVAC team recalled:

Our election officials . . . looked on in disbelief. The computer called for an Eisenhower victory by an overwhelming landslide. The odds by which he would win exceeded the two digits allowed in the program; thus the printout showed 00 to 1 instead of 100 to 1. The officials put their heads together and said "We can't let this go out. The risk is too great." It was beyond their comprehension that with so few votes counted the machine could predict with such a degree of certainty that the odds would be greater than 100 to 1.

The UNIVAC operators quickly adjusted the parameters of the program, so that it would produce more believable results. At 9:15, the UNIVAC's first broadcast prediction anticipated an Eisenhower win with the more modest odds of 8 to 7. But as the night wore on, it became clear that a landslide was indeed developing. A UNIVAC spokesman later appeared on television to come clean and admit to the suppression of UNIVAC's original prediction. The final result was a win for Eisenhower against Stevenson of 442 to 89 electoral votes—within a whisker of UNIVAC's original prediction of 438 to 93.

Of course, UNIVAC's programmers and Remington Rand's managers kicked themselves for holding back the initial prediction, but they could hardly have invented a more convincing demonstration of the apparent infallibility of a computer. Or, rather, of a UNIVAC—for UNIVAC was rapidly becoming the generic name for a computer. The appearance of the UNIVAC on election night was a pivotal moment in computer history. Before that date, while some people had heard about computers, very few had actually seen one; after it, the general public had been introduced to computers and had seen at least a mock-up of one. And that computer was called a UNIVAC, not an IBM.

IBM's Big Push

In fact, IBM had begun to see the writing on the wall several months before UNIVAC's election night stunt. It was the launch of the Census Bureau UNIVAC eighteen months earlier that had galvanized IBM into action. According to Watson junior, he first learned of the completion of the UNIVAC from the chief of IBM's Washington office while they were having a routine meeting. It came to Watson as a bolt from the blue:

I thought, "My God, here we are trying to build Defense Calculators, while UNIVAC is smart enough to start taking all the civilian business away!" I was terrified.

I came back to New York in the late afternoon and called a meeting that stretched long into the night. There wasn't a single solitary soul in IBM who grasped even a hundredth of the potential the computer had. We couldn't visualize it. But the one thing we could understand was that we were losing business. Some of our engineers already had a fledgling effort under way to design a computer for commercial applications. We decided to turn this into a major push to counter UNIVAC.

There was a great deal of ego at the top of IBM, and Watson junior was creating his own mythology. It was absurd to say, as Watson did, that "There wasn't a single solitary soul in IBM who grasped . . . the potential the computer had." As we have seen, IBM had, in the five or six years since the war, developed a superb organizational capability in electronics and computers, and even had two serious data-processing computers under arrested development (the Tape Processing Machine and the Magnetic Drum Calculator). It was a consequence of Watson's autocratic management style that he had—essentially on a hunch—promoted the Defense Calculator to the detriment of the two data-processing computers.

However, one advantage of Watson's autocratic style was that once he had decided that IBM had to move into data-processing computers in a big way—and had convinced his father—it could be made to happen by decree. As of the spring of 1951, IBM had three major computer projects running full steam ahead: the Defense Calculator, the Tape Processing Machine, and the low-cost Magnetic Drum Calculator. These would later emerge as Models 701, 702, and 650. Watson recalled: "People at IBM invented the term 'panic mode' to describe the way we operated: there were moments when we thought we were aboard the *Titanic*." The development cost of the three concurrent projects was enormous, and large numbers of R&D workers had to be recruited. A year later, 35 percent of IBM's laboratory staff was working on electronics, 150 of them on the Defense Calculator alone.

The Defense Calculator, renamed Model 701, was the farthest along of the projects. Originally expected to rent for $8,000 per month, the machine had quickly taken ten orders. In early 1952, however, it became clear that the machine would have to rent for about twice that amount—equivalent to a purchase price of about $1 million, much the same as the UNIVAC was turning out to cost. Customers who had written initial letters of intent for the machine had to be advised of the new rental, but fortunately for IBM most of the orders held up at the revised price. This ready acceptance of the new price reflected the greater maturity and knowledge of computer buyers in 1952: Large, fast scientific computers were going to cost about $1 million, regardless of who made them.

Surprisingly, many conservatives within IBM still argued against developing the Tape Processing Machine, maintaining as late as 1953 that conventional punched-card machines would be cheaper and more cost-effective. However, this failed to take account of the fact that computers were hot news, and business magazines were buzzing with stories about electronic brains for industry and commerce. Cost-effectiveness was no longer the only reason, or even the most important reason, for a business to buy a computer. A January 1952 article in *Fortune* magazine titled "Office Robots" caught the mood exactly:

> The longest-haired computer theorists, impatient of half measures, will not be satisfied until a maximum of the human element and paper recording is eliminated. No intervening clerical operators. No bookkeepers. No punched cards. No paper files. In the utility billing problem, for instance, meter readings would come automatically by wire into the input organs of the central office's electronic accounting and information processing machine, which, when called on, would compare these readings with customers' accounts in its huge memory storage, make all computations, and return the new results to storage while printing out the monthly bills.

Of course, this journalistic hyperbole was little short of science fiction, but Watson and IBM's sales organization realized by this time that IBM had to be seen as an innovator and had to develop a data-processing computer, whether or not it made money.

The first Model 701 scientific computer came off the production line in December 1952, and IBM derived the maximum publicity it could. In a dramatic out-with-the-old-in-with-the-new gesture, designed in part to draw attention away from the UNIVAC, the 701 was installed in the IBM showroom in its New York headquarters—in the place that had been occupied by the SSEC, which was now unceremoniously dismantled.

IBM's first real data-processing computer, the Model 702—based on the TPM—was announced in September 1953. By the following June, IBM had taken fifty orders. However, deliveries did not begin until early 1955, four years after delivery of the first UNIVAC. During the eighteen months between the announcement of the 702 and the first deliveries, it was only Remington Rand's inability to deliver more UNIVACs that enabled IBM to protect its order book.

Although the IBM 702 Electronic Data Processing Machine (EDPM) had superficially much the same specification as the UNIVAC, in the underlying engineering there were major differences that enabled IBM's product quickly to outdistance the UNIVAC in the marketplace. Instead of the mercury delay lines chosen for the UNIVAC, IBM decided to license the Williams Tube technology

developed for the Manchester University computer in England. Both were difficult and temperamental technologies, but the Williams Tube memory was probably more reliable and certainly twice as fast. The magnetic-tape systems on the
700 series were far superior to those of the UNIVAC, which were never completely satisfactory and required constant maintenance. Developing a successful
magnetic-tape system was much more of a mechanical-engineering problem
than an electronic one, and IBM's outstanding capability in electromechanical
engineering enabled it to make a system that was much faster and more reliable.

Another advantage over the UNIVAC was that IBM's computers were modular in construction—that is, they consisted of a series of "boxes" that could be
linked together on site. In addition to making shipment easier, modular construction led to greater flexibility in the size and specification of a machine. By
contrast, the UNIVAC was such a monolith that just moving it from the factory
to a customer's site was a major operation. Unlike UNIVAC, IBM set sizes of
components to fit into standard elevators. Indeed, the first UNIVAC for the
Census Bureau spent its first several months being run in the Philadelphia plant
rather than risk moving it. Finally, IBM's greatest advantage was its reputation
as a service-oriented vendor. Recognizing the importance of training from the
very beginning, it organized programming courses for users and established
field-engineering teams that provided a level of customer service superior to
that offered by any other vendor.

In fact, neither the 701 nor the 702 proved to be quite the "UNIVAC-beater"
that IBM had hoped. For both machines, the Williams Tube memories proved
troublesome, making them unreliable by IBM's normal high standards. By 1953,
however, a new technology known as core memory had been identified. Several
groups worked independently on core memories, but the most important development was the work of Jay Forrester at MIT (see chapter 7). Although it was
still a laboratory prototype at this time, IBM mounted a crash research program
to develop core memory into a reliable product. More reliable and faster core-
memory versions of the 701 and 702 were announced in 1954 as the Model 704
scientific computer and the Model 705 EDPM.

According to Watson junior, 1955 was the turning point for IBM. That year,
orders for IBM's 700 series computers exceeded those for UNIVACs. As late as
August 1955, UNIVACs had outsold 700s by about 30 to 24. But a year later,
there were 66 IBM 700 series installations in place compared with 46 UNIVAC;
and there were, respectively, 193 to 65 machines on order.

Yet it was not the large-scale 700 series that secured IBM's leadership of the
industry, but the low-cost Magnetic Drum Calculator. The machine was announced as the Model 650 in 1953 and rented for $3,250 a month (the equiva-

lent of a purchase price of about $200,000). Although only a quarter of the cost of a 701, it was still a very expensive machine—costing up to twice as much as comparable machines from other makers. However, it succeeded in the market by virtue of its superior engineering, reliability, and software. Eventually 2,000 of these machines were delivered, yielding revenues far exceeding those of the entire 700 series.

As Watson junior observed: "While our giant, million-dollar 700 series got the publicity, the 650 became computing's 'Model T.'" With an astute understanding of marketing, IBM placed many 650s in universities and colleges, offering machines with up to a 60 percent discount provided courses were established in computing. The effect was to create a generation of programmers and computer scientists nurtured on IBM 650s, and a trained workforce for IBM's products. It was a good example of IBM's mastery of marketing, which was in many ways more important than mastery of technology.

With the success of the 650, IBM rapidly began to eclipse Remington Rand's UNIVAC division. Even so, since the launch of the first UNIVAC in 1951, Remington Rand had not been sitting idly by, but had been actively expanding its computer division. In 1952 it had acquired the small Minneapolis-based start-up, Engineering Research Associates, which developed Remington Rand's first scientific computer, the UNIVAC 1103. In mid-1955 Remington Rand merged with Sperry Gyroscope—and was renamed Sperry Rand—to further develop its technology. But in 1954 or 1955 it had already begun to slip behind IBM. Part of the reason for this was a failure in Remington Rand's marketing: The company had been nervous that computers might lower sales of its traditional punched-card equipment and had therefore failed to integrate its punched-card and computer sales staff, often missing sales opportunities. Other reasons for Remington Rand's decline were infighting between its Philadelphia and Minneapolis factories and its unwillingness to invest heavily enough in computer software and marketing. While UNIVAC was still the public's idea of a computer, corporate decision makers were beginning to favor IBM. In the memorable phrase of the computer pundit Herb Grosch, in losing its early lead to IBM, Remington Rand "snatched defeat from the jaws of victory."

The Computer Race

Although the story of IBM versus UNIVAC is a fair metaphor for the progress of the computer industry in the 1950s, the picture was much more complicated. The scene was one of a massive shakeout toward the end of the 1950s.

In the early 1950s the electronics and control manufacturers had been able to participate in the computer business by making machines for science and engineering, if not profitably, then with tolerable losses. By the end of the decade, however, as the computer was transformed into a business machine manufactured and sold in high volumes, these companies were all faced with the same decision: Whether to exit the business or to spend the enormous sums necessary to develop peripherals and applications software, and pay for the marketing costs, to compete with IBM. Of the major firms, only three—RCA, GE, and Honeywell—chose to commit the resources to stay in the race. The remainder, including such established companies as Philco and Raytheon, fell by the wayside.

However, by far the most vulnerable firms were the small computer start-ups, of whom very few survived into the 1960s. They included firms such as the Computer Research Company (CRC—a spin-off of Northrop Aircraft Corporation), Datamatic (a joint venture of Honeywell and Raytheon), Electrodata, the Control Data Corporation (CDC), and the Digital Equipment Corporation (DEC). Of these, only two—CDC and DEC—retained their independence and developed into major firms. And of these two, only CDC had become a major player in mainframe computers by the end of the 1950s. DEC would survive, but it was not until it found a niche in minicomputers that it became a major player (see chapter 9). After old-fashioned bankruptcy, the next most common fate of the small computer firms was to be absorbed by one of the office-machine companies—just as EMCC and ERA had been absorbed by Remington Rand.

Apart from IBM, none of the office-machine firms entered the postwar world with much in the way of electronics experience, nor any interest in making computers. Like IBM, the office-machine companies had prudent doubts about the market for business computers and initially did nothing more than put electronics into their existing products in an evolutionary way. NCR, for example, devised a machine called the "Post-Tronic," which was an electronic enhancement of its ordinary accounting machines supplied to banks. The Post-Tronic, however, was a brilliant success—bigger than almost any computer of its era. The machine created a sensation when it was announced in 1956 at the American Bankers Association meeting. It went on to do $100 million worth of business, and "became an important factor in making NCR's heavy investment for computer development possible." However, to make the final leap into computer production, NCR acquired CRC in 1953 and launched a series of business computers in 1956.

Burroughs also tried to develop its own computer division, and in 1948 hired Irven Travis, then research director of the Moore School, to run its Paoli, Pennsylvania, computer development laboratory. Burroughs initially spent several million dollars a year on electronics research and development, a decision that "represented courage and foresight, because in the 1946–1948 period, our revenue averaged less than $100 million a year and our net profit was as low as $1.9 million in 1946." Despite this investment, Burroughs failed to develop a successful data-processing computer, and so in 1956 acquired the Electrodata Company, a start-up that had already developed a successful business computer, the Datatron. Burroughs paid $20.5 million for Electrodata, an enormous sum at the time, but with the combination of Electrodata's technology and Burroughs's sales force and customer base, it was soon able to secure an acceptable position in the computer company league.

Several other office-machine companies tried to enter the computer race in the 1950s, by either internal development or acquisition. These included Monroe, Underwood, Friden, and Royal, but none of their computer operations survived into the 1960s. Thus, by the end of the 1950s the major players in the computer industry consisted of IBM and a handful of also-rans: Sperry Rand, Burroughs, NCR, RCA, Honeywell, GE, and CDC. Soon, journalists would call them IBM and the seven dwarves.

Despite the gains of the 1950s in establishing a computer industry, as the decade drew to a close, the computer race had scarcely begun. In 1959, IBM was still deriving 65 percent of its income from its punched-card machines in the United States, and in the rest of the world the figure was 90 percent. The 1950s had been a dramatic period of growth for IBM, with its workforce increasing from 30,000 to nearly 100,000, and its revenue growing more than fivefold, from $266 million to $1,613 million. In the next five years, IBM would come to utterly dominate the computer market. The engine that would power that growth was a new computer, the IBM 1401.

6

:: THE MATURING OF THE MAINFRAME: THE RISE AND FALL OF IBM

AS LATE as 1960 IBM was still primarily a punched-card machine supplier. It was not until 1962 that computer sales equaled those of its traditional punched-card products. But by the end of the decade, its punched-card machine sales were essentially vestigial.

While IBM was making this transformation in its product line in the 1960s, it was also growing at the rapid rate of 15 to 20 percent a year and soon achieved a domination of the computer market that was historically unparalleled in any other major industry. From annual sales of $1.8 billion and a head count of 104,000 in 1960, it had rocketed to sales of $7.2 billion and 259,000 employees by the end of the decade. Its market share was over 70 percent in 1960, a position it was able to sustain and even exceed throughout the decade.

The Breakthrough Model 1401

The turning point for IBM was the announcement of the Model 1401 computer in October 1959. The 1401 was not so much a computer as a computer *system*. Most of IBM's competitors were obsessed with designing central processing units, or "CPUs," and tended to neglect the system as a whole. There were plenty of technical people in IBM similarly fixated on the architecture of computers: "Processors were the rage," one IBM technical manager recalled. "Processors were getting the resources, and everybody that came into development had a new way to design a processor." The technocrats at IBM, however, were overridden by the marketing managers, who recognized that customers were more interested in the solution to business problems than in the technical merits of

117

competing computer designs. As a result, IBM engineers were forced into taking a total-system view of computers—that is, a holistic approach that required them to "take into account programming, customer transition, field service, training, spare parts, logistics, etc." This was the philosophy that brought success to the IBM 650 in the 1950s and would soon bring it to the Model 1401. It is, in fact, what most differentiated IBM from its major competitors.

The origin of the 1401 was the need to create a transistorized follow-up to the tube-based Model 650 Magnetic Drum Calculator. By 1958, 800 Model 650s had been delivered—more than all the computers produced by all the other main-frame manufacturers put together. Even so, the 650 had failed to make big in-roads into IBM's several thousand traditional punched-card accounting machine installations. There were a number of sound reasons why most of IBM's cus-tomers were still resisting the computer and continuing to cling to traditional ac-counting machines. Foremost was cost: For the $3,250 monthly rental of a 650, a customer could rent an impressive array of punched-card machines that did just as good a job or better. Second, although the 650 was probably the most reliable of any computer on the market, it was a tube-based machine that was fundamen-tally far less reliable than an electromechanical accounting machine. Third, the would-be owner of a 650 was faced with the problem of hiring programming staff to develop applications software, since IBM provided little programming support at this time. This was a luxury that only the largest and most adventur-ous corporations could afford. Finally, most of the "peripherals" for the 650—card readers and punches, printers, and so on—were lackluster products derived from existing punched-card machines. Hence, for most business users, the 650 was in practice little more than a glorified punched-card machine. It offered few real advantages but many disadvantages. Thus the IBM 407 accounting machine remained IBM's most important product right up to the end of the 1950s.

During 1958 the specification of the 1401 began to take shape, and it was heavily influenced by IBM's experience with the 650. Above all, the new ma-chine would have to be cheaper, faster, and more reliable than the 650. This was perhaps the easiest part of the specification to achieve since it was effectively guaranteed merely by the substitution of improved electronics technology: transistors for tubes and core memory for the magnetic drum of the 650, which would produce an order-of-magnitude improvement in speed and reliability. Next, the machine needed peripherals that would give it decisive advantages over the 650 and electric accounting machines: new card readers and punches, printers, and magnetic-tape units. By far the most important new peripheral under development was a high-speed printer, capable of 600 lines per minute.

IBM also had to find a technological solution to the programming problem. The essential challenge was how to get punched-card-oriented business analysts to be able to write programs without a huge investment in retraining them and without companies having to hire a new temperamental breed of programmer. The solution that IBM offered was a new programming system called Report Program Generator (RPG). It was especially designed so that people familiar with wiring up plugboards on accounting machines could, with a day or two of training, start to write their own business applications using familiar notations and techniques. The success of RPG exceeded all expectations, and it became one of the most used programming systems in the world. However, few RPG programmers realized the origins of the language or were aware that some of its more arcane features arose from the requirement to mimic the logic of a punched-card machine.

Not every customer wanted to be involved in writing programs however, even with RPG; some preferred to have applications software developed for them by IBM—for applications such as payroll, invoicing, stock control, production planning, and other common business functions. Because of its punched-card machine heritage, IBM knew these business procedures intimately and could develop software that could be used with very little modification by any medium-sized company. IBM developed entire program suites for the industries that it served most extensively, such as insurance, banking, retailing, and manufacturing. These application programs were very expensive to develop, but because IBM had such a dominant position in the market it could afford to "give" the software away, recouping the development costs over tens or hundreds of customers. Because IBM's office-machine rivals—Sperry Rand, Burroughs, and NCR—did not have nearly so many customers, they tended to develop software for the one or two industries in which they had traditional strengths rather than trying to compete with IBM across the board.

The IBM 1401 was announced in October 1959, and the first systems were delivered early in 1960 for upward of $2,500 per month rental (the equivalent of a purchase price of $150,000). This was not much more than the cost of a medium-sized punched-card installation. IBM had originally projected that it would deliver around 1,000 systems. This turned out to be a spectacular underestimate, for 12,000 systems were eventually produced.

How did IBM get its forecast so wrong? The 1401 was certainly an excellent computer, but the reasons for its success had very little to do with the fact that it was a computer. Instead, the decisive factor was the new type 1403 "chain" printer that IBM supplied with the system. The printer used a new technology by which slugs of type were linked in a horizontal chain that rotated rapidly and were hit on the fly by a series of hydraulically actuated hammers. The printer achieved a speed of 600 lines per minute, compared with the 150 lines per minute of the popular 407 accounting machine that was based on a prewar printing technology. Thus, for the cost of a couple of 407s, the 1401 provided the printing capacity of four standard accounting machines; and the flexibility of a stored-program computer came, as it were, for free. It was an unanticipated motive for IBM's customers to enter the computer age, but no less real for that.

For a while, IBM was the victim of its own success, as customer after customer decided to turn in old-fashioned accounting machines and replace them with a computer. In this decision they were aided and abetted by IBM's industrial designers, who excelled themselves by echoing the modernity and appeal of the new computer age: out went the round-cornered steel-gray punched-card machines, and in came the square-cornered light-blue computer cabinets. For a firm with less sophisticated financial controls and less powerful borrowing capabilities than IBM, coping with the flood of discarded rental equipment would have been a problem. But as a reporter in *Fortune* noted:

> [F]ew companies in U.S. business history have shown such growth in revenues and such constantly flourishing earnings. I.B.M. stock, of course, is *the* standard example of sensational growth sustained over many years. As everybody has heard, anyone who bought 100 shares in 1914 (cost, $2,750), and put up another $3,614 for rights along the way, would own an investment worth $2,500,000 today.

With a reputation like that, the unpredicted success of a new product was a problem that IBM's investors were willing to live with. As more and more 1401s found their way into the nation's offices in their powder-blue livery, IBM earned a sinister new name: Big Blue.

IBM and the Seven Dwarves

Meanwhile, IBM's success was creating a difficult environment for its competitors. By 1960 the mainframe computer industry had already been whittled down to just IBM and seven others. Of all the mainframe suppliers, Sperry Rand had suffered the biggest reverse, consolidating a decline that had begun well before the launch of the 1401. Despite being the pioneer of the industry, it had never made a profit in computers and was gaining a reputation bordering on derision. Another *Fortune* writer who eulogized IBM noted:

> The least threatening of I.B.M.'s seven major rivals . . . has been Sperry Rand's Univac Division, the successor to Remington Rand's computer activities. Few enterprises have ever turned out so excellent a product and managed it so ineptly. Univac came in too late with good models, and not at all with others; and its salesmanship and software were hardly to be mentioned in the same breath with I.B.M.'s.

As a result "the upper ranks of other computer companies are studded with ex-Univac people who left in disillusionment." In 1962 Sperry Rand brought in an aggressive new manager from ITT, Louis T. Rader, who helped UNIVAC address its deficiencies. But despite making a technologically successful entry into computer systems for airline reservations, Rader was soon forced to admit, "It doesn't do much good to build a better mousetrap if the other guy selling mousetraps has five times as many salesmen."

In 1963 UNIVAC turned the corner and started to break even at last. But the machine that brought profits, the UNIVAC 1004, was not a computer at all but a transistorized accounting machine, targeted at its existing punched-card machine users. Even with the improving outlook, UNIVAC still had only a 12 percent market share and a dismal one-sixth as many customers as IBM. This was little better than the market share that Remington Rand had secured with punched-card machines in the 1930s. It seemed that the company would be forever second.

IBM's other business-machine rivals, Burroughs and NCR, were a long way behind IBM and even UNIVAC, both having market shares of around 3 percent—only a quarter of that of UNIVAC. In both cases, their strategy was safety first: protect their existing customer base, which was predominantly in the banking and retail sectors, respectively. They provided computers to enable their existing customers to move ahead with the new technology in an evolutionary way, as they felt the need to switch from electromechanical to electronic machines.

Doing much better than any of the old-line office-machine companies, the fastest-rising star in the computer scene of the early 1960s was Control Data Corporation, the start-up founded in 1957 by the entrepreneur William Norris. Control Data had evolved a successful strategy by manufacturing mainframes with a better price performance than those of IBM—by using superior electronics technology and by selling them into the sophisticated scientific and other markets where the IBM sales force had not so great an impact. By 1963 Control Data had risen to third place in the industry, not far behind UNIVAC.

The only other important survivors in the mainframe industry of the early 1960s were the computer divisions of the electronics and control giants—RCA, Honeywell, and General Electric. All three had decided to make the necessary investment to compete with IBM. In every case this was a realistic possibility only because they rivaled IBM in power and size. The difference, however, was that for all of them the computer was not a core business, but an attempt to enter a new and unfamiliar market. Each had plans to make major product announcements toward the end of 1963.

Honeywell was planning to launch its Model 200 computer that would be compatible with the IBM 1401 and would be able to run the same software, and therefore IBM's existing customers would be an easy target. Honeywell's technical director, a wily ex-UNIVAC engineer named J. Chuan Chu, had reasoned that IBM was sufficiently nervous of its vulnerability to antitrust litigation to be "too smart not to let us take 10 per cent of the business." Inside RCA, General David Sarnoff, its legendary chief executive officer, had committed the company to spending huge sums to develop a range of IBM-compatible computers and was in the process of making licensing agreements with overseas companies to manufacture the computers in the rest of the world. Finally, General Electric also had plans to announce a range of three computers—one small, one medium-sized, and one large—in late 1963.

Revolution, Not Evolution: System/360

Although in marketing terms the IBM 1401 had been an outstanding success, inside IBM there was a mishmash of incompatible product lines that threatened its dominance of the industry. Many IBM insiders felt that this problem could be resolved only by producing a compatible range of computers, all having the same architecture and running the same software.

In 1960 IBM was producing no less than seven different computer models—some machines for scientific users, others for data-processing customers; some large machines, some small, and some in between. In manufacturing terms IBM was getting precious little benefit from its enormous scale. By simultaneously manufacturing seven different computer models, IBM was being run almost as a federation of small companies instead of an integrated whole. Each computer model had a dedicated marketing force trained to sell into the niche that computer occupied, but these specialized sales forces were unable to move easily from one machine or market niche to another. Each computer model required a dedicated production line and its own specialized electronic components. Indeed, IBM had an inventory of no less than 2,500 different circuit modules for its computers. Peripherals also presented a problem, since hundreds of peripheral controllers were required so that any peripheral could be attached to any processor.

All this has to be compared with IBM's punched-card products, where a single range of machines (the 400 series accounting machines) satisfied all its customers. The resulting rationalization of production processes and standardization of components had reduced manufacturing costs to an extent that IBM had no effective competition in punched-card machines at all.

The biggest problem, however, was not in hardware but in software. Because the number of software packages IBM offered to its customers was constantly increasing, the proliferation of computer models created a nasty gearing effect: Given m different computer models, each requiring n different software packages, a total of $m \times n$ programs had to be developed and supported. This was a combinatorial explosion that threatened to overwhelm IBM at some point in the not-too-distant future. It was a problem rather like the one facing the Parliament of the European Community at the same time. Because the six countries of the Community used four different languages—French, German, Italian, and Dutch—it was necessary to provide twelve different translation services (French to German, Italian, and Dutch; German to French, Italian, and Dutch; and so on). As the Community grew to nine members in 1973, and eventually to twelve, the Parliament had to accept a core of three languages (English, French, and German) so that only six translation services had to be provided instead of the dozens that would otherwise be necessary. Of course, if members of the European Parliament had spoken Esperanto, then no translators would have been needed at all. For IBM, a compatible range promised to be the electronic Esperanto that would contain the software problem.

Just as great a problem was that of the software written by IBM's customers. Because computers were so narrowly targeted at a specific market niche, it was not possible for a company to expand its computer system in size by more than a factor of about two without changing to a different computer model. If this was done, then all the user's applications had to be reprogrammed. This often caused horrendous organizational disruption during the changeover period. Indeed, reprogramming could be more costly than the new computer itself. As IBM appreciated only too well, once a company had decided to switch computer models, it could look at computers from all manufacturers—not just those made by IBM.

All these factors suggested that IBM would sooner rather than later have to embrace the compatible-family concept. But in IBM's case this would be particularly challenging technically because of the wide spectrum of its customers in terms of size and applications. A compatible series would have to satisfy all of IBM's existing customers, from the very smallest to the very largest, as well as both its scientific and commercial customers. And the new machine would have to be compatible throughout the product range, so that programs written on one machine would execute on any other—more slowly on the small machines, certainly, but they would have to run without any reprogramming at all.

The decision to produce a compatible family was not so clear-cut as it appears in hindsight, and there was a great deal of agonizing at IBM. For one thing, it was by no means clear that compatibility to the extent proposed was technically feasible. And even if it were, it was feared that the cost of achieving compatibility might add so much to the cost of each machine that it would not be competitive in the marketplace. Another complication was that there were factions within IBM that favored consolidating its current success by building on its existing machines. The 1401 faction, for example, wanted to make more powerful versions of the machine. These people thought it was sheer madness for IBM to think of abandoning its most successful product ever. If IBM dropped the 1401, they argued, it would leave thousands of disaffected users open to competitors. Another faction inside IBM favored, and had partially designed, a new range, to be known as the 8000 series, to replace IBM's large 7000 series machines.

But it was the software problem that was to determine product strategy, and by late 1960 the tide was beginning to turn toward the radical solution. Not one of IBM's computers could run the programs of another, and if IBM was to introduce more computer models, then, as a top executive stated, "we are going to wind up with chaos, even more chaos than we have today." For the next several months planners and engineers began to explore the technical and managerial problems of specifying the new range and of coordinating the fifteen to twenty computer development groups within IBM to achieve it. Progress was slow, not least because those involved in the discussions had other responsibilities or preferred other solutions—and the compatible-family concept was still no more than a possibility that might or might not see the light of day.

To resolve the compatible-family debate rapidly, in October 1961 T. Vincent Learson, Tom Watson Jr.'s second-in-command at IBM, established the SPREAD task group, consisting of IBM's thirteen most senior engineering, software, and marketing managers. SPREAD was a contrived acronym that stood for Systems,

Programming, Review, Engineering, And Development, but which was really meant to connote the broad scope of the challenge in establishing an overall plan for IBM's future data-processing products. Progress was slow and, after a month, Learson, an archetypal IBM vice president, became impatient for a year's end decision. In early November he banished the entire task group to a motel in Connecticut, where it would not be distracted by day-to-day concerns, with "orders not to come back until they had agreed."

The eighty-page SPREAD Report, dated 28 December 1961, was completed on virtually the last working day of the year. It recommended the creation of a so-called New Product Line that was to consist of a range of compatible computers to replace all of IBM's existing computers. On 4 January the SPREAD Report was presented to Watson Jr., Learson, and the rest of IBM's top management. The report was breathtaking in its scope—and in the expense of making it a reality. For example, it was estimated that software alone would cost $125 million—at a time when IBM spent just $10 million a year on all its programming activities. Learson recalled that there was little enthusiasm for the New Product Line at the meeting:

> The problem was, they thought it was too grandiose. . . . The job just looked too big to the marketing people, the financial people, and the engineers. Everyone recognized it was a gigantic task that would mean all our resources were tied up in one project—and we knew that for a long time we wouldn't be getting anything out of it.

But Watson and Learson recognized that to carry on in the same old way was even more dangerous. They closed the meeting with the words, "All right, we'll do it."

Implementation of the New Product Line got under way in the spring of 1962. Considerable emphasis was placed on security. For example, the project was blandly known as NPL (for New Product Line), and each of the five planned processors had a misleading code number—101, 250, 315, 400, and 501—which gave no hint of a unified product line and in some cases was identical to the model numbers of competitive machines from other manufacturers; even if the code numbers leaked, they would merely confuse the competition.

The New Product Line was one of the largest civilian R&D projects ever undertaken. Until the early 1980s, when the company began to loosen up somewhat, the story of its development was shrouded in secrecy. Only one writer succeeded in getting past the barrier of IBM's press corps, the *Fortune* journalist

Tom Wise. He coined the phrase "IBM's $5 billion gamble" and wrote "not even the Manhattan Project which produced the atomic bomb in World War II cost so much"; this sounded like hyperbole at the time, but Wise's estimate was about right. Wise reported that one of the senior managers "was only half joking when he said: 'We called this project "You bet your company."'" It is said that IBM rather liked the swashbuckling image that Wise conveyed, but when his articles went on to recount in detail the chaotic and irrational decision-making processes at IBM, Watson Jr. was livid and "issued a memorandum suggesting that the appearance of the piece should serve as a lesson to everyone in the company to remain uncommunicative and present a unified front to the public, keeping internal differences behind closed doors."

The logistics of the research program were awesome. Of the five planned computer models, the three largest were to be developed at IBM's main design facility in Poughkeepsie, New York, the smallest in its Endicott facility in upstate New York, and the fifth machine in its Hursley Development Laboratories in England. Simply keeping the machine designs compatible between the geographically separate design groups was a major problem. Extensive telecommunications facilities were used to keep the development groups coordinated, including two permanently leased transatlantic lines—an unprecedented expense in civilian R&D projects at the time. In New York hundreds and eventually thousands of programmers worked on the software for the New Product Line.

All told, direct research costs were around $500 million. But ten times as much again was needed for development—to tool up the factories, retrain marketing staff, and re-equip field engineers. One of the major costs was to build up a capability in semiconductor manufacturing, in which IBM had previously been weak. Tom Watson Jr., who had to cajole his board of directors into sanctioning the expenditure, recalled:

> Ordinary plants in those days cost about forty dollars per square foot. In the integrated circuit plant, which had to be kept dust free and looked more like a surgical ward than a factory floor, the cost was over one hundred and fifty dollars. I could hardly believe the bills that were coming through, and I wasn't the only one who was shocked. The board gave me a terrible time about the costs. "Are you really sure you need all this?" they'd say. "Have you gotten competitive bids? We don't want these factories to be luxurious."

This investment put IBM into the position of being the world's largest manufacturer of semiconductors.

By late 1963, with development at full momentum, top management began to turn its thoughts to the product launch. The compatible range had now been dubbed System/360, a name "betokening all points of the compass" and suggesting the universal applicability of the machines. The announcement strategy was fraught with difficulty. One option was to make a big splash by announcing the entire series at once, but this carried the risk that customers would cancel their orders for existing products and IBM would be left with nothing to sell until the new range came on stream. A safer and more conventional strategy would be to announce one machine at a time over a period of a couple of years. This would enable the switch from the old machines to the new to be achieved much more gently, and it was, in effect, how IBM had managed the transition from its old punched-card machines to computers in the 1950s.

However, all internal debate about the announcement strategy effectively ceased in December 1963 when Honeywell announced its Model 200 computer. The Honeywell 200 was the first machine to challenge IBM by aggressively using the concept of IBM compatibility. The Honeywell computer was compatible with IBM's Model 1401, but by using more up-to-date semiconductor technology it was able to achieve a price-performance superiority of as much as a factor of four. Because the machine was compatible with the 1401, it was possible for one of IBM's customers to return an existing rented machine to IBM and acquire a more powerful model from Honeywell for the same cost—or a machine of the same power for a lesser cost. Honeywell 200s could run IBM programs without reprogramming and, using a provocatively named "liberator" program, could speed up existing 1401 programs to make full use of the Honeywell 200's power.

It was a brilliant strategy and a spectacular success—Honeywell took 400 orders in the first week following the announcement, more than it had taken in the previous eight years of its computer operation's existence. The arrival of the Honeywell 200 produced a noticeable falloff in IBM 1401 orders, and some existing users began to return their 1401s to switch over to the new Honeywell machine. Inside IBM it was feared that as many as three-quarters of its 1401 users would be tempted by the Honeywell machine.

Despite all the planning done on System/360, IBM was still not irrevocably committed to the New Product Line, and the arrival of the Honeywell 200 caused all the marketing plans to be worked over once again. At this, the eleventh hour, there seemed to be two alternatives: To carry on with the launch of the entire System/360 range and hope that the blaze of publicity and the implied obsolescence of the 1401 would eclipse the Honeywell 200; or to abandon

System/360 and launch a souped-up version of the 1401—the 1401S— that was already under development and that would be competitive with the Honeywell machine.

On 18–19 March a final make-or-break decision was made in a protracted risk-assessment session with Watson Jr., IBM's president Al Williams, and the thirty top IBM executives:

> At the end of the risk assessment meeting, Watson seemed satisfied that all the objections to the 360 had been met. Al Williams, who had been presiding, stood up before the group, asked if there were any last dissents, and then, getting no response, dramatically intoned, "Going . . . going . . . gone!"

Activity in IBM now reached fever pitch. Three weeks later the entire System/360 product line was launched. IBM staged press conferences in sixty-three cities in the United States and fourteen foreign countries on the same day. In New York a chartered train conveyed two hundred reporters from Grand Central Station to IBM's Poughkeepsie plant, and the visitors were shown into a large display hall where "six new computers and 44 new peripherals stretched before their eyes." Tom Watson Jr.—grayhaired, but a youthful not-quite fifty, and somewhat Kennedyesque—took center stage to make "the most important product announcement in company history," IBM's third-generation computer, System/360.

The computer industry and computer users were stunned by the scale of the announcement. While an announcement from IBM had long been expected, its tight security had been extremely effective, so that outsiders were taken aback by the decision to replace the entire product line. Tom Wise, the *Fortune* reporter, caught the mood exactly when he wrote:

> The new System/360 was intended to obsolete virtually all other existing computers. . . . It was roughly as though General Motors had decided to scrap its existing makes and models and offer in their place one new line of cars, covering the entire spectrum of demand, with a radically redesigned engine and an exotic fuel.

This "most crucial and portentous—as well as perhaps the riskiest—business judgment of recent times" paid off handsomely. Literally thousands of orders for System/360 flooded in, far exceeding IBM's ability to deliver; in the first two years of production it was able to satisfy less than half of the 9,000 orders on its books. To meet this burgeoning demand, IBM hired more marketing and production staff and opened new manufacturing plants. Thus in the three years following the System/360 launch, IBM's sales and leasing revenue soared to over $5 billion, and its employee head count increased by 50 percent, taking it to nearly a quarter of a million workers.

System/360 has been called "the computer that IBM made, that made IBM." The company did not know it at the time, but System/360 was to be its engine of growth for the next thirty years. In this respect, the new series was to be no more and no less important to IBM than the classic 400 series accounting machines that had fueled its growth from the early 1930s until the computer revolution of the early 1960s.

The Dwarves Fight Back

In the weeks that followed the System/360 announcement, IBM's competitors began to formulate their responses. System/360 had dramatically reshaped the industry around the concept of a software compatible computer range. Although by 1964 the compatible-range concept was being explored by most manufacturers, IBM's announcement effectively forced the issue.

Despite IBM's dominance of the mainframe computer industry, it was still possible to compete with the computer giant simply by producing better products, which was by no means impossible. In the mythology of computer history, System/360 has often been viewed as a stunning technical achievement and "one of this country's greatest industrial innovations"—a view that IBM naturally did all it could to foster. But in technological terms, System/360 was no more than competent. The idea of a compatible range was not revolutionary at all, but was well understood throughout the industry; and IBM's implementation of the concept was conservative and pedestrian. For example, the proprietary electronics technology that IBM had chosen to use, known as Solid Logic Technology (SLT), was halfway between the discrete transistors used in second-generation computers and the true integrated circuits of later machines. There were good, risk-averse decisions underlying IBM's choice, but to call System/360 the "third generation" was no more than a marketing slogan that stretched reality.

Perhaps the most serious design flaw in System/360 was its failure to support time-sharing—the fastest-growing market for computers—which enabled a machine to be used simultaneously by many users. Another problem was that the entire software development program for System/360 was little short of a fiasco (as we shall see in chapter 8) in which thousands of programmers consumed over $100 million to produce software that had many shortcomings. IBM was not so cocooned from the real world that it was unaware of what poor value it had derived from the $500 million System/360 research costs. As he caught wind of the new computer's failings, Tom Watson Jr. personally demanded an engineering audit, which fully confirmed his suspicion that IBM's engineering was mediocre.

But as always, technology was secondary to marketing in IBM's success—and there was no question of trying to compete with IBM in marketing. Hence, to compete with IBM at all, the seven-dwarf competitors had to aim at one of IBM's technical weak spots.

The first and most confrontational response was to tackle IBM head-on by making a 360-compatible computer with superior price performance—just as Honeywell had done with its 1401-compatible Model 200. RCA, one of the few companies with sufficient financial and technical resources to make the attempt, decided on this strategy. Indeed, two years before the System/360 announcement, RCA had committed $50 million to produce a range of IBM-compatible computers. RCA had deliberately kept its plans flexible so that it would be able to copy whatever architecture and instruction code IBM finally adopted, which it believed "would become a standard code for America and probably for the rest of the world."

Right up to the launch of System/360 in April 1964, RCA product planners had no inkling either of the date of the announcement or of the eventual shape of System/360. There was no industrial espionage, and it was only on the day of the announcement that they got their first glimpse of the new range. Within a week they had decided to make the RCA range fully System/360-compatible, and working from IBM's publicly available manuals "started to copy the machine from the skin in." In order to compete with IBM, RCA needed a 10 or 15 percent advantage in price/performance; this it planned to achieve by using its world-leading capability in electronic components to produce a series that used fully integrated circuits in place of the SLT technology used by IBM. This would make the RCA machines—dollar for dollar—smaller, cheaper, and faster than IBM's. RCA announced its range as Spectra 70 in December 1964—some eight months after the System/ 360 launch.

RCA was the only mainframe company to go IBM-compatible in a big way. The RCA strategy was seen as very risky because it was bound to be very difficult to achieve a superior price performance to IBM, given IBM's massive economies of scale. IBM compatibility also put the competitor in the position of having to follow slavishly every enhancement that IBM made to its computer range. Even worse, IBM might drop System/360 altogether one day.

Hence, the second and less risky strategy to compete with System/360 was product differentiation—that is, to develop a range of computers that were software compatible with one another but not compatible with System/360. This was what Honeywell did. Taking its highly successful Model 200, it extended the machine up and down the range to make a smaller machine, the Model 120, and four larger machines, the 1200, 2200, 4200, and 8200, which were announced between June 1964 and February 1965. This was a difficult decision for Honeywell, since it could be seen as nailing its flag to the mast of an obsolete architecture. However, Honeywell had the modest ambition of capturing just 10 percent of IBM's existing 1401 customer base—about 1,000 installations in all—which it comfortably did. The same strategy, of producing non-IBM-compatible ranges, was also adopted by Burroughs with its 500 series, announced in 1966, and by NCR with its Century Series in 1968.

The third strategy to compete with IBM was to aim for niche markets that were not well satisfied by IBM and in which the supplier had some particular competitive advantage. Control Data, for example, recognized that the System/360 range included no really big machines, and so decided to give up manufacturing regular mainframes altogether and make only giant number-crunching computers designed primarily for government and defense research agencies. This strategy was very successful, and soon Control Data became the world's third-largest computer maker. Similarly, General Electric recognized IBM's weakness in time-sharing, and—collaborating with computer scientists at Dartmouth College and MIT— quickly became the world leader in time-shared computer systems (see chapter 9). In other cases, manufacturers capitalized on their existing software and applications experience. Thus UNIVAC continued to compete successfully with IBM in airline reservation systems, without any major machine announcements. Again, even though their computer products were lackluster, Burroughs and NCR were able to build on their special relationship with bankers and retailers, which dated from the turn of the century, when they had been selling adding machines and cash registers. It was the old business of selling solutions, not problems. With these strategies and tactics, all seven dwarves survived into the 1970s—just.

By the end of the 1960s, IBM appeared invincible. It had about three-quarters of the market for mainframe computers around the world. Its seven mainframe competitors each had between 2 and 5 percent of the market; some were losing money heavily, and none of them was more than marginally profitable. IBM's extraordinarily high profits, estimated at 25 percent of its sales, enabled the other companies to shelter under its "price umbrella." It was often said that the seven dwarves existed only by the grace of IBM, and if it were ever to fold its price umbrella, they would all get wet—or drown. This is what happened in the first computer recession of 1970–71. As the journalists put it at the time, IBM sneezed and the industry caught a cold. Both RCA and General Electric, which had problems in their core electronics and electrical businesses associated with the global economic downturn of the early 1970s, decided to give up the struggle and terminate their in-the-red computer divisions. Their customer bases were bought up by Sperry Rand and Honeywell, respectively, which helped them gain market share at relatively little cost. From then on the industry was no longer characterized as IBM and the seven dwarves, but as IBM and the BUNCH— standing for Burroughs, UNIVAC, NCR, Control Data, and Honeywell.

The Future System

Meanwhile, inside IBM, thought was being given to the future of the System/360. In the short run, it was decided to enhance the existing range by incorporating some straightforward technological advances to maintain its competitiveness. In the long run, however, it intended to replace System/360 with a completely new architecture, code-named FS for Future System.

In June 1970 the evolutionary successor to the 360 range was announced as System/370. The new series offered improved price/performance through updated electronic technology: True integrated circuits were used in place of the SLT modules of System/360, and magnetic core storage was replaced with semiconductor memory. The architecture was also enhanced to support more effectively time-sharing and communications-based on-line computing. The technique of "virtual memory" was also introduced, using a combination of software and hardware to increase the amount of working memory in the computer (and therefore permit the execution of much larger programs). IBM secured great publicity value from the "innovation" of virtual memory, although the idea had been pioneered nearly a decade earlier at Manchester University in England. As always, IBM's publicity machine was stronger than its technology.

While the life of the 360–370 line was being extended by these modest improvements, IBM's main R&D resources were focused on the much more challenging Future System. The new range was planned to be ready for market in the second half of the 1970s. As with System/360 a decade earlier, there was no unanimity within IBM about the wisdom of migrating to a new computer architecture. IBM's top-level Corporate Technical Committee argued, however, that sooner or later the company would have to make a decision. In effect, the committee was proposing a rerun of the System/360 story:

> To have competitive leadership will require bringing together new technology for logic and memory, new architecture, new programming and new input/output devices as well as CPUs. This is a major undertaking and a task equal to, if not larger than, our change to System/360.

In mid-1971, a task group was established to determine the broad capabilities of IBM's Future System, just as ten years earlier the SPREAD task group had evolved the design of System/360.

The concepts behind FS were very advanced, as least as advanced as System/360 had been in its day. One overarching theme was the realization that only one-third of a customer's costs were for hardware—the rest was for programming and services. Thus a primary goal of FS was to reduce software costs through a completely new design that would "displace the 370 line with a system of novel architecture, much as System/360 had displaced its predecessors." In the spring of 1971 the task group's recommendation to replace System/360 was accepted by management, and it planned to announce the new range in June 1975—allowing some forty-five months for research and development.

Unfortunately, the FS program had nothing like the tight focus of the System/360 program of the early 1960s. Deadline after deadline slipped by, and after three years of work the project stubbornly remained far from completion. This slippage may have been partly the result of democratic dithering in IBM following the retirement of the autocratic Tom Watson Jr. as CEO in 1971. But there were formidable technical problems too:

> People involved with the project agree that it failed largely because of the vastness of the undertaking. Some suggest the organization and leadership were inadequate. Others focus on the conflicts and uncertainties surrounding the issue of direct implementation versus the three-level architecture. Many more believe the FS objectives were themselves the problem, simply too far ahead of available technologies.

FS was indeed a remarkably advanced architecture, and this entailed its own problems. Even today there is no major computer range that has the functionality of FS as it was proposed in the early 1970s.

But more than anything, it was the growing importance of the existing software investment of IBM and its users that made a new architecture less feasible by the day and sealed the fate of FS. The prospect of rewriting all that software was awesome:

> In terms of sheer effort, it was estimated that all the programmers in IBM would be kept busy until 1978 implementing FS; none would be available to work on incremental improvements to System/370. This was unacceptable in the competitive marketplace in which System/370 had to survive until FS was ready.

For customers, the prospect of rewriting their code was equally unacceptable. System/360 and its successor had become an industry standard.

In February 1975, IBM pulled the plug and decided to drop FS and instead continue to evolve the 360–370 line. The design work on FS had involved thousands of staff years of effort and was estimated to have cost $100 million, making it "the most expensive development-effort failure in IBM's history."

IBM had been hoist with its own petard. For the decade of the 1960s, its growth had been assisted by the "software lock-in" that prevented its customers from defecting to competitors. Now IBM was locked in, too. Of course, the problem of software compatibility that forced IBM to remain wedded to its 360 architecture also forced other computer manufacturers to stay with their computer designs.

Thus by the mid-1970s, the mainframe had fully matured. Today, all of the world's mainframe designs are about thirty to forty years old; and much of the software running on these mainframes is equally old. The maturing of the mainframe has produced one of the great technological contradictions of the twentieth century: As semiconductor components improve in size and speed by a factor of two every year or two, they are used to build fossilized computer designs and run software that is groaning with age. The world's mainframes and the software that runs on them have become like the aging sewers beneath British city streets, laid in the Victorian era. Not a very exciting infrastructure, and largely unseen—but they work, and life would be totally different without them.

The Decline of the IBM Empire

By the mid-1970s information technology was one of the world's major industries, ranking with automobiles and petroleum. Of every $120 of goods produced in the United States, one dollar was on computer systems. It was, of course, inevitable that IBM could not sustain a three-quarter market share indefinitely, as manufacturing globalized and the market began to fragment. For example, the advent of low cost integrated circuits saw the arrival in the early 1970s of minicomputer manufacturers such as DEC, Data General, and Hewlett-Packard (see chapter 9). IBM competed vigorously and successfully with these new companies, but it never secured the same kind of dominance in minicomputers that it had in mainframes.

Thus by 1976, although IBM still held two-thirds of the world market for mainframes, its overall share of the global computer market had been reduced to 50 percent. By 1985 it was down to 25 percent. But this reflected only a broadening of the market. IBM's growth and profitability scarcely faltered through the 1970s and the first half of the 1980s. It grew at an annual rate of 15 to 20 percent a year, peaking with sales of $50 billion and a workforce over 400,000 in 1985.

Yet within a period of five years IBM was to suffer the most dramatic fall from grace in corporate history, recording ever more slender profits, until in January 1993 it reported the largest loss ever by a private company. How did this happen?

For many years IBM's domination of the computer industry was attributed to a variety of factors: its managerial competence; its technological excellence; its formidable marketing organization; its monopolistic business practices; and the leadership exerted by the Watsons. Yet when IBM fell into decline, very little had changed. Its technology and marketing were as good in the 1990s as at any time in its past; and although it had only a 20 percent market share, it exceeded its nearest competitor in size by a factor of five; and even though the Watsons were no longer at the helm of the company, IBM's leaders were still among the most respected in the world. This suggests that there was something in the environment beyond IBM that was responsible for its rise and fall.

IBM's malaise was frequently attributed to the rise of the personal computer, but the facts do not support this contention. The scene was set for IBM's downfall in the mid-1970s, with the commodification of the mainframe—that is, by the mainframe becoming an article that could be produced by any competent manufacturer. This had the inevitable result of reducing the generous profit margins IBM had enjoyed for sixty years.

Today, with the hindsight of historical perspective, IBM's early domination of the computer market can be seen as largely a fortuitous inheritance. For twenty years, from the mid-1950s to the mid-1970s, IBM was blessed with a unique combination of organizational capabilities that equipped it perfectly for the mainframe computer market. While it fit this environmental niche, IBM prospered wonderfully; but once that niche changed, IBM's strategies were no longer appropriate. Its dominance of the mainframe had very little to do with electronics but everything to do with the capabilities it brought from its origins as a manufacturer of punched-card accounting machines.

IBM's key competencies were electromechanical manufacturing, marketing, and a service orientation—all geared to the delivery of integrated data-processing systems. Electromechanical manufacturing was enormously important to IBM—much more so than electronics. As we have seen, what made the Model 1401 so successful was not electronics but its path-breaking 600-lines-per-minute printer; and it was IBM's electromechanical peripherals, such as magnetic tapes and disk drives, that consolidated this position. Again, IBM's sales organization was second to none, and it remained as true in Thomas Watson Jr.'s time as when his father had said in the 1930s that "IBM's products are sold, not bought." (It is noteworthy that the only two mainframe suppliers to pull out in the early 1970s were RCA and General Electric. These were electronics and control-equipment manufacturers with no preexisting sales organization for marketing office equipment.) Customer service always had the highest priority in IBM. The firm was celebrated for responding to a malfunctioning computer system with a dedicated engineering task force until the problem was fixed.

However, it was IBM's background as a supplier of integrated data-processing systems that provided its unique advantage. As Frank Cary, a future president of IBM, put it in June 1964, shortly after the launch of System/360, "We don't sell a product . . . we sell solutions to problems." This attitude was very different from that of IBM's office-machine competitors. Before they became computer manufacturers, Sperry Rand, Burroughs, and NCR had supplied products such as typewriters, accounting machines, and cash registers. These products were generally supplied as "office appliances" that could be dropped into any office, leaving the underlying information system unchanged. Thus a secretary's work was facilitated by a typewriter, a bookkeeper's work was speeded up by an accounting machine, and a teller in a store was assisted by a cash register. None of these products required the office-machine supplier to have more than a superficial knowledge of the underlying business.

By contrast, when IBM sold a punched-card machine system, the business's entire information system had to be reconfigured around the machinery. This key difference between IBM and its competitors persisted right into the computer age. Thus, when a company used equipment from one of IBM's office-machine competitors, it was all too likely to acquire a problem rather than a solution. Often the computer and its software were no more than a set of tools with which to fashion a solution, instead of the solution itself.

Only IBM guaranteed a complete solution to business problems, and an IBM salesman was all too likely to remind a data-processing manager that no one ever got fired for hiring equipment from IBM. This was a patronizing attitude that came close to condescension, and often resulted in a love-hate relationship between IBM and its customers. In the mid-1970s a great change began to occur in the computer industry, which was fully realized in the 1980s; 1973 marked the first year that more than 100,000 computers were in use worldwide. But the growth slowed, and in 1978 there were only 7,000 more computers in use than there had been five years earlier. The mainframe had matured and become a commodity, while IBM's expertise in systems integration was increasingly taken over by software. As a result, computer users had far less need of IBM to hold their hand, and it was inevitable that they would begin to drift to other, less intimidating and less expensive suppliers.

Part Three

INNOVATION AND EXPANSION

:: REAL TIME:
REAPING THE WHIRLWIND

MOST OF the computer systems installed in commerce and government during the 1950s and 1960s were used unimaginatively. They were simply the electronic equivalents of the punched-card accounting machines they replaced. While computers were more glamorous than punched-card machines, giving an aura of modernity, they were not necessarily more efficient or cheaper. For many organizations, computers did not make any real difference; data-processing operations would have gone on much the same without them. Where computers made a critical difference was in real-time systems.

A *real-time* system is one that responds to external messages in "real time," which usually means within seconds, although it can be much faster. No previous office technology had been able to achieve this speed, and the emergence of real-time computers had the potential for transforming business practice.

Jay Forrester and Project Whirlwind

The technology of real-time computing did not come from the computer industry. Rather, the computer industry appropriated a technology first developed for a military purpose in MIT's Project Whirlwind, a project that developed out of a contract to design an "aircraft trainer" during World War II.

An aircraft trainer was a system developed during the war to speed up the training of pilots to fly new aircraft. Instead of learning the controls and handling characteristics of an airplane by actually flying one, a pilot would use a simulator, or trainer: a mock-up of a plane's cockpit, with its controls and instruments, attached to a control system. When the pilot "flew" the simulator, the electromechanical control system fed the appropriate data to the aircraft's instruments, and a series of mechanical actuators simulated the pitch and yaw of an aircraft. Sufficient realism was attained to train a pilot far more cheaply and safely than using a real airplane.

A limitation was that a different simulator had to be built for each type of aircraft. In the fall of 1943 the Special Devices Division of the U.S. Bureau of Aeronautics established a project to develop a universal flight trainer that would be suitable for simulating any type of aircraft. This could be achieved only with a much more sophisticated computing system. A contract for an initial feasibility study was given to the Servomechanisms Laboratory at MIT.

MIT's Servomechanisms Laboratory was the leading center for the design of military computing systems in the United States. It had been established in 1941 as part of the war effort to develop sophisticated electromechanical control and computing devices for use in fire control (that is, gun aiming), bomb aiming, automatic aircraft stabilizers, and so on. By 1944 the laboratory had a staff of about a hundred people working on various projects for the military. An assistant director of the laboratory, Jay W. Forrester, was placed in charge of the project. Then only twenty-six, Forrester showed an outstanding talent for the design and development of electromechanical control equipment.

Forrester was a very resilient individual who, for the next decade, drove his research project forward in the face of a hostile military establishment that justifiably balked at its cost. Many of the early computer projects ran far over their original cost and time estimates, but none so much as the Whirlwind. It changed its objectives midway; it took eight years to develop instead of the projected two; and it cost $8 million instead of the $200,000 originally estimated. In a more conservative and rational research environment, the project might have been terminated. The project did manage to survive and had an impact that eventually did justify the expenditure. Apart from the SAGE and SABRE projects (to be described later), Project Whirlwind established real-time computing and helped prepare the foundations of the "Route 128" computer industry in Massachusetts. In 1944, however, all this lay in the unforeseeable future.

For the first few months of 1945, Forrester worked virtually alone on the project, drawing up the specification of the new-style trainer. He envisioned a simulator with an aircraft cockpit with conventional instruments and controls harnessed to an analog computing system that, by means of a hydraulic transmission system, would feed back the correct operational behavior of the plane being modeled. The cockpit and instruments appeared to be relatively straightforward; and although the problem of the analog computer was much more challenging, it did not appear insurmountable.

On the basis of Forrester's initial study, MIT proposed in May 1945 that the trainer project be undertaken by the Servomechanisms Laboratory, at a cost of some $875,000 over a period of eighteen months. In the context of the war with Japan—whose end was still four months in the future—$875,000 and eighteen months seemed like a sound investment, and MIT was given the go-ahead. During the following months, Forrester began to recruit a team of mechanical and electrical engineers. He recruited a high-flying MIT graduate student, Robert E. Everett (later to become president of the MITRE Corporation) to serve as his assistant. Over the next decade, while Forrester retained overall administrative and technical leadership, Everett assumed day-to-day responsibilities as Forrester became increasingly absorbed with financial negotiations.

As the engineering team built up, Forrester put into motion the development of the more straightforward mechanical elements of the aircraft trainer, but the problem of the computing element remained stubbornly intractable. However much he looked at it, it was clear that an analog computer would not be nearly fast enough to operate the trainer in real time. The solution to the problem lay in a remarkable convergence of computing and control technologies that was to change the nature of Forrester's project and ultimately become a major sector of postwar technology.

Seeking a solution to the control problem, Forrester found time in his busy schedule to survey the computing field. In the summer of 1945 he learned for the first time about digital computers from a fellow MIT graduate student, Perry Crawford. Crawford was perhaps the first to appreciate real-time digital control systems, long predating the development of practical digital computers. As early as 1942 he had submitted a master's thesis on *Automatic Control by Arithmetic Operations*, which discussed the use of digital techniques for automatic control. Up to that date, control systems had always been implemented using mechanical or electromechanical analog technologies. These techniques were widely used in the hundreds of control devices built into World War II weapons. The characteristic shared by all these devices was that they used some physical analog to model a mathematical quantity.

For example, an array of discs and gear wheels could be used to perform the mechanical integration of the differential equations of motion, or a "shaped resistor" could be used to model a mathematical function electrically. By contrast, in a digital computer, mathematical variables were represented by numbers pure and simple, stored in digital form, just as they were stored in a cash register or in a desktop calculating machine. Crawford was the first person to fully appreciate that such a general-purpose digital computer would be potentially faster and more flexible than a dedicated analog computer. All this Crawford explained to Forrester while the two were deep in conversation on the steps in front of the MIT building on 77 Massachusetts Avenue. As Forrester later put it, Crawford's remarks "turned on a light in [my] mind." The course of Project Whirlwind was set.

With the end of the war in August 1945, the digital computation techniques that had been developed in secret during the war rapidly began to diffuse into the civilian scientific arena. In October, the National Defense Research Committee organized a Conference on Advanced Computational Techniques at MIT, which was attended by both Forrester and Crawford. It was here that Forrester learned in detail of the "Pennsylvania Technique" of electronic digital calculation used in the ENIAC at the University of Pennsylvania, and in its successor, the EDVAC. When the Moore School ran its summer school the following year, Crawford gave a lecture on the real-time application of digital computers. Everett was formally enrolled as a student, and he or one of the other engineers attended every session.

Meanwhile, the trainer project had acquired its own momentum, even though the computing problem remained unresolved. As was typical of university projects, Forrester staffed the project with young graduate engineers and gave them considerable independence. This produced an excellent esprit de corps within the laboratory, but outside the result was a reputation for arrogance and an impression that the project was a "gold-plated boondoggle." One of Forrester's Young Turks, Kenneth Olsen—subsequently founder and president of the Digital Equipment Corporation—later remarked, "Oh we were cocky! We were going to show everybody! And we did."

By early 1946 Forrester had become wholly convinced that the way forward was to follow the digital path. Even though this would result in a major cost escalation, he believed the creation of a general-purpose computer for the aircraft trainer would permit the "solution of many scientific and engineering problems other than those associated with aircraft flight."

In March, Forrester made a revised proposal to the Special Devices Division of the Bureau of Aeronautics requesting that his contract be renegotiated to allow for the development of a full-scale digital computer in addition to the original aircraft trainer. It was estimated that the cost of the computer would be $1.9 million, compared with a little under half a million dollars for the trainer. This was a threefold cost escalation. But by laying meticulous technical groundwork and cultivating the support of his superiors at MIT, Forrester succeeded in selling the revised program—now formally named Project Whirlwind—to the Bureau.

Despite Forrester's confidence, it must be recalled that the construction of the first stored-program digital computer lay three years in the future, and the Whirlwind computer would not emerge until two years after that. During 1946, therefore, it was a matter of learning as much as possible about the new world of digital computing and finding or inventing the basic technology. Forrester's whole laboratory now became reoriented to the design and development of Whirlwind. There was a total of about a hundred staff members, working in ten groups and costing $100,000 a month to support. One group under Everett was working on the "block diagrams" (or computer architecture, as we would now call it), another group was at work on electronic digital circuits, two groups were engaged in storage systems, another group worked on mathematical problems, and so on.

As did everyone leading an early computer project, Forrester soon discovered that the single most difficult problem was the creation of a reliable storage technology. By comparison, building the computer's arithmetic processor was relatively easy, even though it would contain about 3,000 tubes. The most promising storage technology was the mercury delay-line memory, on which many other centers were working. But this system would be far too slow for the Whirlwind. While mathematical laboratories were working on computers that would run at speeds of 1,000 or 10,000 operations per second, the MIT machine, if it was successfully to fulfill its real-time function, needed to run 10 or 100 times faster—on the order of 100,000 operations per second.

Forrester therefore tried another promising technology—the electrostatic storage tube—which was a modified form of TV tube in which each bit of information was stored as an electrical charge. Unlike the mercury delay-line memory, which used sonic pulses, the electrostatic memory could read and write information at electronic speeds of a few microseconds. In 1946, however, electrostatic storage was no more than a blue-sky idea, and a long and expensive research project lay ahead. Moreover, storage was just one component—albeit the most difficult—of the whole system. As the magnitude of the digital computer project began to dawn, the original flight trainer aspect receded into the background so that the manpower could be deployed on the more pressing computer problem. As the official historians of Project Whirlwind later put it, "the tail had passed through and beyond the point of wagging the dog and had become the dog."

The years 1946 and 1947 passed fruitfully in developing and testing the Whirlwind computer's basic components. While great strides were being made—not just in Forrester's group but in computer groups around the world—no one had yet demonstrated a working computer, so that the technical progress seemed un-dramatic from the outside. In the case of Project Whirlwind, the most dramatic characteristic remained Forrester's capacity for spending large sums of money.

In the outside world, Forrester's cozy relationship with the Bureau of Aeronautics was coming to an end. In the immediate postwar period, military funding had been reorganized, and Project Whirlwind now came under the remit of the Office of Naval Research (ONR), which took over the Special Devices Unit of the Bureau of Aeronautics. While expensive crash projects had been normal in time of war, a new value-for-money attitude had accompanied the shrinking defense budgets of the postwar environment. From late 1947 Forrester found himself under increasing critical scrutiny from the ONR, and several independent inquiries were commissioned into the cost-effectiveness of Project Whirlwind and its scientific and engineering competence. Forrester tended to shield his team from these political and financial problems, as he "saw no reason to allocate time for his engineers to stand wringing their hands."

Fortunately Forrester was blessed with one key ally in the ONR: Perry Crawford. Crawford had joined the ONR as a technical adviser in October 1945, and his vision for real-time computing was undiminished. Indeed, if anything, it had widened even beyond that held by Forrester. Crawford foresaw a day when real-time computing would control not just military systems but whole sectors of the civilian economy, such as air traffic control. In October 1947 Crawford, Forrester, and Everett coauthored a report on *Information Systems of Inter-connected Digital Computers*, which presented the concept of a combat information control system using computers. (Later this background was to help Forrester secure a role for Whirlwind in the SAGE computer-based air-defense system.)

Meanwhile, Project Whirlwind was consuming 20 percent of ONR's total research budget, and its cost was far out of line with the other computer projects it was funding. The ONR could not understand why, when all the other computer groups were building computers with teams of a dozen or so engineers and budgets of half a million dollars or less, Forrester needed a staff of more than one hundred and a budget approaching $3 million. Moreover, the other computer groups included people a good deal more prestigious than those at Project Whirlwind—such as von Neumann's group at the Institute for Advanced Study in Princeton.

There were two reasons for Whirlwind's much higher costs. First, it needed to be an order of magnitude faster than any other computer to fulfill its real-time function. Second, it needed to be engineered for reliability because break-

downs that would cause no more than a minor irritation in a routine mathematical computation were unacceptable in a real-time environment. Unfortunately, the people in ONR who were responsible for computer policy were
primarily mathematicians; while they appreciated the utility of computers for
mathematical work, they had not yet been convinced of their potential for control applications. By the summer of 1948 the ONR's confidence in Forrester and
Project Whirlwind was ebbing fast. At the same time, Forrester's main ally,
Perry Crawford, with his "fearless and imaginative jumps into the future," was
under a cloud and had been moved to another assignment. This left Forrester to
handle the ONR as best he could.

A major confrontation occurred in the fall of 1948 when Forrester requested
a renewal of his ONR contract at a rate of $122,000 a month. This was met by
a counteroffer of $78,000 a month—representing a budget cut of 38 percent.
Forrester's reaction was outrage that the project should be trimmed when he believed it was on the threshold of success. Moreover he now saw Project Whirlwind simply as a stepping stone to a major military information system (the
original aircraft trainer having long been forgotten), which would take perhaps
fifteen years to develop at a cost of $2 billion. Forrester argued that Whirlwind
was a national computer development of such self-evident importance that the
costs, however great, were immaterial. On this occasion bravado and brinkman-
ship won the day, but a further crisis was not long in coming.

Meanwhile Forrester may have half-convinced the ONR that Whirlwind was
on the threshold of success, but by no means had all the technical issues been
resolved—especially the storage problem. The electrostatic storage tubes had
proved expensive and their storage capacity was disappointing. Forrester, always on the lookout for an alternative memory technology, was struck by an
advertisement he saw for a new magnetic ceramic known as Deltamax. In June
1949 he made an entry in his notebook sketching out the idea for a new kind of
memory, and he quietly set to work. A former graduate student in the laboratory later recalled: "Jay took a bunch of stuff and went off in a corner of the lab
by himself. Nobody knew what he was doing, and he showed no inclination to
tell us. All we knew was that for about five or six months or so, he was spending
a lot of time off by himself working on something." The following fall Forrester
handed the whole project over to a young graduate student named Bill Papian
and awaited results.

By late 1949 inside ONR confidence in Project Whirlwind had reached its
lowest point. The head of the computer branch, C. V. L. Smith, was dismayed by
Forrester's high-flown vision of national military information systems, which

he found both "fantastic" and "appalling." Another critic found the Whirlwind to be "unsound on the mathematical side" and "grossly over-complicated technically." A high-level national committee found the Whirlwind to be lacking "a suitable end-use" and that if the ONR could not find one, then "further expenditure on the machine should be stopped." Smith had all the evidence he needed, and in 1950 he reviewed Project Whirlwind's budget with Forrester. Rather than providing the budget of $1.15 million that Forrester requested for fiscal 1951, he offered an amount in the region of a quarter of a million dollars to bring Whirlwind into a working state, and then the ONR could wash its hands of the whole project. Fortunately as one door shut another had already begun to open.

The SAGE Defense System

The new door started to open, quite independently, in August 1949 when U.S. intelligence revealed that the Russians had exploded a nuclear bomb and that they possessed bomber aircraft capable of carrying such a weapon over the North Pole and into the United States. America's defenses against such an attack were feeble. The existing air-defense network, which was essentially left over from World War II, consisted of a few large radar stations where information was manually processed and relayed to a centralized command-and-control center. The greatest weakness of the system was its limited ability to process and make operational use of the information it gathered. There were communication difficulties at all levels of the system, and there were gaps in radar coverage, so that, for example, no early warning could be provided of an attack coming from over the North Pole.

The heightened Cold War tension of the fall of 1949 provided the catalyst to get the nation's air-defense system reviewed by the Air Force's Scientific Advisory Board. A key member of this board was George E. Valley, a professor of physics at MIT. Valley made his own informal review of the air-defense system that confirmed its inadequacy. In November 1949 he proposed that the Board create a subcommittee with the remit of recommending the best solution to the problem of air defense.

This committee, the Air Defense System Engineering Committee (ADSEC), came into being the following month. Known informally as the Valley Committee after its originator and chairman, it made its first report in 1950, describing the existing system as "lame, purblind, and idiot-like." It recommended upgrading the entire air-defense system to include updated interceptor aircraft, ground-to-air missiles, and anti-aircraft artillery, together with improved radar coverage and computer-based command-and-control centers.

At the beginning of 1950 Valley was actively seeking computer technology to form the heart of a new-style command-and-control center when an MIT colleague pointed him in the direction of Forrester and Project Whirlwind. Valley later recalled: "I remembered having heard about a huge analog computer that had been started years before; it was an enterprise that I had carefully ignored, and I had been unaware that the Forrester project had been transformed into a digital computer." As Valley asked around, the reports he received about Project Whirlwind "differed only in their degree of negativity."

Valley decided to find out for himself and made a visit to Project Whirlwind, where Forrester and Everett showed him all he wanted to see. By a stroke of good fortune, the Whirlwind had just started to run its first test programs—stored in the bare twenty-seven words of electrostatic memory that were available. Valley was able to see the machine calculate a freshman mechanics problem and display the solution on the face of a cathode-ray tube. This demonstration, combined with Forrester and Everett's deep knowledge of military information systems, was enough to sway him toward Whirlwind—though in truth there was no other machine to be had.

At the next Project Whirlwind review meeting with the ONR on 6 March 1950, George Valley made an appearance. He stated his belief that the air force would be willing to allocate some $500,000 to the project for fiscal 1951. This independent commendation of the Whirlwind transformed the perception of Forrester's group within the ONR, and the negative criticism began to disappear. Over the next few months further air force monies were found, enabling the project to proceed on the financial terms Forrester had always insisted upon. A few months after, Whirlwind became the major component of a much larger project, Project Lincoln—later the Lincoln Laboratory—established at MIT to conduct the entire research and development program for a computerized national air-defense system. It would take another decade for the resulting air-defense system—the Semi-Automatic Ground Environment (SAGE)—to become fully operational.

Meanwhile, the development of Project Whirlwind continued at full pace. Problems continued to be experienced with building reliable electrostatic storage tubes, but as the core-memory idea had still not borne fruit, they had to press ahead with both possibilities. Input-output devices were commissioned, such as conventional paper-tape readers and punches, and typewriter terminals; another specially designed device displayed words and pictures on the face of a cathode-ray tube. In addition applied research studies were undertaken to process radar data in the Whirlwind; for example, trials were conducted for the computer-controlled interception of alien bombers by guided missiles.

By the spring of 1951 the Whirlwind computer was an operational system. The only major failing was the electrostatic storage tubes, which continued to be unreliable and had a disappointing capacity. The solution to this problem finally came into sight at the end of 1951 when Bill Papian demonstrated a prototype core-memory system. Further development work followed, and in the summer of 1953 the entire unreliable electrostatic memory was replaced by core memory with an access time of 9 microseconds—making Whirlwind by far the fastest computer in the world and also the most reliable.

Papian was not the only person working on core memory—there were at least two other groups working on the idea at the same time—but the "MIT core plane" was the first system to become operational. In five years' time, core memory would replace every other type of computer memory and net MIT several million dollars in patent royalties. The value to the nation of the core-memory spin-off could by itself be said to have justified the cost of the entire Whirlwind project.

With the installation of the core memory and the consequent improvement in speed and reliability, Whirlwind was effectively complete. The research and development phase was over, and now production in quantity for Project SAGE had to begin. Gradually the technology was transferred to IBM, which transformed the prototype Whirlwind into the IBM AN/FSQ-7 computer. In 1956 Forrester left active computer engineering to become an MIT professor of industrial and engineering organization in the Sloan School of Management; he subsequently became well known to the public for his computer-based studies of the global environment, published as *World Dynamics* in 1971.

The SAGE system that was eventually constructed consisted of a network of twenty-three Direction Centers distributed throughout the country. Each Direction Center was responsible for air surveillance and weapons deployment in a single sector—typically a land/sea area of several thousand square miles. At the heart of each Direction Center was an IBM AN/FSQ-7 computer—the production version of Whirlwind. The computer was "duplexed" for reliability—one machine operating live and the other in standby mode. With 49,000 tubes and weighing 250 tons, the computer was the largest ever put into service. Feeding the computer were some one hundred sources of data coming into the Direction Center. These sources included ground-based and ship-based radar installations, weapons and flight bases, on-board missile and airplane radar, and other Direction Centers. Within the Direction Center over one hundred air force personnel monitored and controlled the air defense of their sector. Most of these people sat at consoles that displayed surveillance data on CRT screens—giving the positions and velocities of all aircraft in their area, while suppressing irrelevant information. By interrogating the console, the operator could determine the identity of aircraft and weapons—whether civilian or military, friendly or alien—and make tactical plans based on the use of computer-stored environmental and weather data, flight plans, and weapons characteristics.

The complete SAGE system, which was fully deployed by 1963 at an estimated cost of $8 billion, was an "expensive white elephant" from a strictly military point of view, however. The original purpose of defending the United States from bomber-borne nuclear weapons had been undercut by the arrival of the new technology of Inter-Continental Ballistic Missiles (ICBMs) in the 1960s. The ICBMs made bombers a second-order threat, although still enough of a concern to provide continuing support for the SAGE system until the early 1980s, when it was finally decommissioned.

The real contribution of SAGE was thus not to military defense, but through technological spin-off to civilian computing. An entire subindustry was created as industrial contractors and manufacturers were brought in to develop the basic technologies and implement the hardware, software, and communications. These industrial contractors included IBM, Burroughs, Bell Laboratories, and scores of smaller firms. Many of the technological innovations that were developed for SAGE were quickly diffused throughout the computer industry. For example, the development of printed circuits, core memories, and mass-storage devices—some of the key enabling technologies for the commercial exploitation of computers in the 1960s—was greatly accelerated by SAGE. The use of CRT-based graphical displays created not just a novel technology but a whole culture of interactive computing that the rest of the world did not catch up with until the 1970s. American companies also gained a commanding lead in digital communications technologies and wide-area networks. Finally, the SAGE system provided a training ground for software engineering talent: Some 1,800 programmer years of effort went into the SAGE software, and as a result "the chances [were] reasonably high that on a large data-processing job in the 1970s you would find at least one person who had worked with the SAGE system."

Of the contractors, IBM was undoubtedly the major beneficiary; as Tom Watson Jr. put it, "It was the Cold War that helped IBM make itself the king of the computer business." The SAGE project accounted for half a billion dollars of IBM's income in the 1950s, and at its peak about 7,000 to 8,000 people were working on the project—some 20 percent of IBM's workforce. IBM obtained a lead in processor technology, mass-storage devices, and real-time systems, which it never lost and soon put to commercial use.

SABRE: A Revolution in Airline Reservations

The first major civilian real-time project to draw directly on IBM's know-how and the technology derived from the SAGE project was the SABRE airline reservations system developed for American Airlines. SABRE, and systems like it, transformed airline operations and helped create the modern airline as we know it. To understand the impact of real-time reservations systems on the airlines, one first needs to appreciate how reservations were handled before computers came on the scene.

If during the 1940s or 1950s you had managed to gain access to the reservations office of a major commercial airline, you would have been confronted by an extraordinary sight, one reminiscent of a war operations room. At the center of the room, amid the noisy chattering of teletype machines, about sixty reservations clerks would be dealing with incoming telephone calls at the rate of several thousand a day. At any one moment, each reservations clerk would be responding to one or more of three general types of inquiry from a customer or travel agent: requests for information about flight availability; requests to reserve or cancel seats on a particular flight; and requests to purchase a ticket. To deal with any of these requests, the reservations clerks had to refer to a series of well-lit boards displaying the availability of seats on each flight scheduled to depart over the next few days. For flights further ahead in time, the agent would have to walk across the room to consult a voluminous card file.

If the inquiry resulted in a reservation, a cancellation, or a ticket sale, the details of the transaction would be recorded on a card and placed in an out tray. Every few minutes these cards would be collected and taken to the designated availability-board operator, who would then adjust the inventory of seats available for each flight. Once a ticket was sold and the availability board had been updated, the sales information found its way to the back office, where another forty or so clerks maintained passenger information and issued tickets. The whole process was altogether more complicated if a trip involved connecting flights; these requests came and went out via teletypes to similar reservations offices in other airlines. The day or two before the departure of a flight produced a flurry of activity as unconfirmed passengers were contacted, last-minute bookings and cancellations were made, and a final passenger list was sent to the departure gate at the airport.

Although tens of thousands of reservations were processed in this manner every day, in reservations offices at several different airports, there was almost no mechanization. The reason for this largely manual enterprise—it could have been run on almost identical lines in the 1890s—was that all the existing data-processing technologies, whether punched-card machines or computers, operated in what was known as batch-processing mode. Transactions were first batched in hundreds or thousands, then sorted, then processed, and finally totaled, checked, and dispatched. The primary goal of batch processing was to reduce the cost of each transaction by completing each subtask for all transactions in the batch before the next subtask was begun. As a result the time taken for any individual transaction to get through the system was at least an hour, but more typically half a day. For most businesses, batch processing was a cost-effective solution to the data-processing problem and had been employed since the 1920s and 1930s: It was how banks processed checks, insurance companies issued policies, and utility companies invoiced their customers. As computers replaced punched-card machines, they usually continued to be used in a batch-processing mode. However, batch processing—whether on punched-card machines or computers—could not address the instantaneous processing needs of an airline reservations system.

The fact that airline reservations were done manually was not a problem until the 1950s. Up to that date, air travel was viewed by the general public as a luxury purchase and so was not particularly price-sensitive. Airlines could still turn a profit operating with load factors—that is, the proportion of seats occupied by fare-paying passengers—that were well below capacity.

However, sweeping changes were becoming apparent in the airline market in the mid-1940s. In 1944, when American Airlines first decided to automate its reservations operation, it was approaching the limit of what was possible using purely manual methods. As the volume of business increased and more and more flights were added to the availability boards, the boards became more crowded and the airlines' reservations clerks had to sit farther and farther away from them—sometimes even resorting to field glasses. In an effort to resolve this problem, the airline commissioned an equipment manufacturer, Teleregister, to develop an electromechanical system that would maintain an inventory of the number of seats available on each flight. Using this system, called the Reservisor, a reservations clerk could determine the availability of a seat on a particular flight using a specially designed terminal, eliminating the need for the availability boards.

The Reservisor brought considerable benefits, enabling the airline to cope with 200 extra flights in the first year of operation with twenty fewer clerks. However, a major problem was that the Reservisor was not directly integrated into the overall information system. Once a sale had been recorded by the Reservisor, the passenger and reservation records still had to be adjusted manually. Discrepancies inevitably arose between the manual system and the Reservisor, and it was estimated that about one in twelve reservations was in error. It was said that "most business travelers returning home on a Friday would have their secretaries book at least two return flights." The result of these errors led either to underselling a flight, causing a financial loss, or overbooking it, with consequent damage to customer relations.

Further refinement led to the Magnetronic Reservisor installed at La Guardia Airport in New York in 1952. Capable of storing details of a thousand flights ten days ahead, the key advance of this new system was an arithmetic capability enabling a sales agent in the New York area to sell and cancel seats as well as simply determine availability. Even so, the Reservisor maintained nothing more than a simple count of the number of seats sold and available on each aircraft—passenger record keeping and ticket issuing still had to be done manually. The manual part of the reservations operation continued to expand so that American Airlines was planning for a new reservations office occupying 30,000 square feet, with seating positions for 362 reservations clerks and a capacity for dealing with 45,000 telephone calls a day.

By 1953 American Airlines had reached a crisis. Increased traffic and scheduling complexity were making reservations costs insupportable. This situation was exacerbated by the coming of $5 million jet airplanes. American Airlines was planning to buy 30 Boeing 707s, each with 112 passengers seats, which would reduce the transcontinental flying time from ten to six hours. This was faster than the existing reservations system could transmit messages about a flight, making it pointless to update passenger lists for last-minute additions or no-shows. The airline was also having to respond to increasing competition. The most important way profitability could be increased was by improving load factors.

American Airlines' data-processing problems were thus high on the agenda of its president, C. R. Smith, when by chance he found himself sitting next to Blair Smith (unrelated), one of IBM's senior salespeople, on a flight from Los Angeles to New York in the spring of 1953. Naturally, their conversation quickly turned to the airline reservations problem. Subsequently, Blair Smith was invited to visit the La Guardia reservations office, after which there was continued contact between the engineering and planning staffs of the two companies.

IBM was in fact well aware of the Reservisor system but had not shown any interest when the system was first developed in the 1940s. By 1950, however, Thomas Watson Jr. had begun to assume control of IBM and had determined that it would enter the airlines reservations market as well as similar "tough applications, such as Banks, Department Stores, Railroads, etc." In 1952 the company had hired Perry Crawford away from the Office of Naval Research, and he was to play a key role in early real-time developments; as early as the summer of 1946 Crawford had been publicly advocating the development of control systems for commercial airlines as well as the more obvious military applications. A small joint study group was formed, under the co-leadership of Crawford, with the brief of exploring the development of a computerized real-time reservations system that would capitalize on IBM's SAGE experience.

In June 1954 the joint study team made its initial report. The report noted that although a computerized reservations system was a desirable goal in the long run, it would be some years before cost-effective technology would become available. The report therefore recommended that, in the short term, the existing record-keeping system should be automated with conventional punched-card equipment. Thus an impressive, though soon-to-be-obsolete, new punched-card system and an improved Reservisor were brought into operation during 1955–59. Simultaneously an idealized, integrated, computer-based reservations system was planned, based on likely future developments in technology. This later became known as a "technology-intercept" strategy. The system would be economically feasible only when reliable solid-state computers with core memories became available. Another key requirement was a large random-access disk storage unit, which at that time was a laboratory development rather than a marketable product. Although, in the light of its SAGE experience, IBM was well placed to make informed judgments about emerging technologies and products, a technology-intercept approach would have been too risky without the backup of the parallel development of a system using conventional punched-card machine technology.

At the beginning of 1957 the project was formally established. During the next three years, the IBM–American Airlines team prepared detailed specifications, while IBM technical advisers selected mainframes, disk stores, and telecommunications facilities; and a special low-cost terminal based on an IBM Selectric typewriter was developed for installation at over a thousand agencies.

In 1960 the system was formally named SABRE—a name inspired by a 1960 advertisement for the Buick LeSabre—for which the somewhat contrived acronym Semi-Automatic Business Research Environment was invented. The acronym was later dropped, although the system continued to be known as SABRE, or Saber, as a name suggesting speed and accuracy. The complete system specification was presented to American Airlines' management in the spring of 1960. The total estimated cost was just under $40 million. After agreeing to go ahead, American Airlines' President Smith remarked: "You'd better make those black boxes do the job, because I could buy five or six Boeing 707s for the same capital expenditure."

The system was fully implemented during 1960–63. The project was at the time easily the largest civilian computerization task ever undertaken, involving some 200 technical personnel producing a million lines of program code. The central reservations system, housed at Briarcliff Manor, about thirty miles north of New York City, was based on two IBM 7090 mainframe computers, which were duplexed for reliable operation—one computer live and the other waiting in standby mode. These processors in turn were connected to sixteen disk storage units with a total capacity of 800 million characters—the largest on-line storage system ever assembled up to that time. All told, through over 10,000 miles of leased telecommunication lines, the system networked some 1,100 agents using desktop terminals in fifty cities throughout the United States. The system was capable of handling close to 10 million passenger reservations per year, involving "85,000 daily telephone calls—30,000 daily requests for fare quotations—40,000 daily passenger reservations—30,000 daily queries to and from other airlines and 20,000 daily ticket sales." Most transactions were handled in three seconds.

More important than the volume of transactions, however, the system provided a dramatic improvement in the reservations service. In place of the simple inventory of seats sold and available as provided by the original Reservisor, there was now a constantly updated and instantly accessible passenger name record containing information about the passenger, including telephone contacts, special meal requirements, and hotel and automobile reservations. The system quickly took over not just reservations, but the airline's total operation: flight planning, maintenance reporting, crew scheduling, fuel management, air freight, and most other aspects of its day-to-day operations.

The system was fully operational in 1964, and by the following year it was stated to be amply recouping its investment through improved load factors and customer service. The project had taken ten years to implement. This in itself is a revealing illustration of IBM's mature and well-seasoned engineering approach derived from its SAGE experience. The development may have seemed leisurely from the outside, but an orderly, evolutionary approach was imperative when replacing the information system of such a massive business. In engineering terms it was the difference between a prototype and a product. All told, the system cost a reported $30 million, and although disparagingly known as "the Kids' SAGE" it was a massive project in civilian terms and not without risk.

What was innovation for one airline soon became a competitive necessity for the others. During 1960 and 1961 Delta and Pan American signed up with IBM, and their systems became operational in 1965. Eastern Airlines followed in 1968. Other alliances between computer manufacturers and airlines developed, some of which lacked the conservative approach of the IBM–American Airlines system and resulted in "classic data-processing calamities" that took years to become operational. Overseas, the major international carriers began developing reservations systems in the mid-1960s. By the early 1970s all the major carriers possessed reliable real-time systems and communications networks that had become an essential component of their operations, second in importance only to the airplanes themselves. Perhaps a single statistic gives a hint of the extraordinary developments that were to follow: By 1987, the world's largest airline reservations system, operated by United Airlines, made use of eight powerful IBM 3090 computers that processed over 1,000 internal and external messages in a single *second.*

The airline reservations problem was unusual in that it was largely unautomated in the 1950s and so there was a strong motivation for adopting the new real-time technology. Longer-established and more traditional businesses were much slower to respond.

Historically, the accounting systems of old, established organizations had evolved symbiotically with office machinery. Office machines were originally designed in the last decades of the nineteenth century to mechanize existing accounting methods. But as office routine became increasingly mechanized in the 1920s and 1930s, it was the accounting methods that adapted to what could be done by bookkeeping machinery. As a result most accounting systems became batch-oriented, and financial information became historic and periodic.

To take just one example, the monthly statements most people receive from their banks today, though prepared by a computer, are essentially identical to those produced by punched-card machines in the 1920s. Behind the scenes the procedures for processing checks are equally archaic. This is a classic example of the persistence of the system—the way new technologies generally overlay existing structures without changing the functions of the components of the system or the relationships between them. Thus, although real-time computers and communications became available in the 1970s that could have provided much better customer information and instantaneous check processing, the inertia in the existing banking system prevented these innovations from taking place. It was mainly where there was a novel and competitive opportunity—such as automatic teller machines—that real-time systems were implemented. A similar picture could be seen throughout industry and commerce; many utilities, insurance companies, manufacturers, and government agencies continued to use bureaucratic and accounting procedures that were creaking with age for another twenty years or more.

Real-time systems are more than a high-speed convenience, however. Batch-processing data-processing systems represent an economic inefficiency because they always contain some capital trapped in the system—whether they are cash transactions being processed overnight or buffering stocks held in manufacturing and retailing. When real-time systems have entirely replaced batch systems, not only will businesses be more reactive but significant amounts of productive capital will be released.

The Universal Product Code

Despite the leisurely progress of many older firms and bureaucracies, from the mid-1960s real-time computing gradually took hold, often out of view of the general public. A notable exception is the universal product code with which we all interact at the supermarket. The universal product code (UPC) is the bar code now almost universally found on food products manufactured or packaged for retail. The UPC was built on the foundations of the modern food manufacturing and distribution industries that emerged in the 1920s and 1930s. The most important early innovation of the supermarket was the labor-saving concept of moving customers past the goods in a store (that is, self-service) instead of having a salesperson bring the goods to the customer.

Although the first supermarket, the Piggly Wiggly store in Memphis, Tennessee, opened in 1916, it took twenty years for supermarkets to become established. The recession of the early 1930s provided the right economic conditions for the early "cheapy" supermarkets to thrive in the eastern United States—operations such as King Kullen and Big Bear. By setting prices some 5 percent below those of traditional grocery stores, they turned over enormous volumes while maintaining profit margins of 3 or 4 percent of sales—more than double the industry average. In 1936, when reliable statistics begin, there were just 600 supermarkets in the United States, but huge structural changes were afoot. In that year A&P, the nation's largest food retailer, had only twenty supermarkets, compared with 14,000 traditional stores. Apart from being a necessary response to competition from the cheapies, a switch by A&P to supermarket operations promised an improvement in operating margins through economies of scale. As each supermarket was opened, several small grocery stores closed. By 1941, the A&P had 1,646 supermarkets, while the number of its traditional stores had declined to a little over 4,000. In the whole country there were 8,000 supermarkets. In parallel with these developments in supermarket retailing, food manufacturing saw national-branded goods, which had begun in a small way in the 1890s, come to dominate the food industry with their promotion in newspapers and through radio advertising in the 1930s and 1940s. Supermarket growth resumed after the war, with over 15,000 supermarkets opened by 1955, and over 30,000 by 1965.

By the end of the 1960s the U.S. food industry had become dominated by the supermarket, which accounted for three-quarters of the total grocery business. But, as is characteristic of a mature industry, the generous profit margins of the boom years were being eroded. By about 1970 productivity had become stagnant and operating margins were as low as 1 percent, a level the industry had not experienced since the 1920s. Moreover, in a period of high inflation, food prices had become a politically sensitive issue, and this put more pressure on the supermarkets to reduce costs by raising productivity. The most obvious area for productivity improvement was the checkout, which accounted for almost a quarter of operating costs and often caused bottlenecks that exasperated customers.

The need to improve these "front-end" operations coincided with the development of a new generation of electronic cash registers, known as point-of-sale terminals, that offered the possibility of automatic data capture for pricing and stock control. The new point-of-sale terminals required the application of product codes to merchandise; the information gathered at the checkout could then be used to maintain inventories and determine the cost of goods. But it was not generally suitable for supermarkets: The large number of product lines offered (10,000 to 20,000) and the low average value of items made the individual application of product codes too expensive.

Attempts were made by a couple of individual grocery chains to implement product coding, which were operationally successful but not cost-effective. The structure of the food industry was such that product coding would be economical only if individual firms acted cooperatively. Even greater benefits would come if manufacturers, wholesalers, and retailers all used the same, universal product code. This would allow the whole of the food industry to harmonize their computer hardware and software development, in effect becoming one integrated operation. This implied an extraordinary degree of cooperation among all these organizations, and the implementation of the UPC was thus as much a political achievement as a technical one.

In 1969 the National Association of Food Chains appointed the management consultants McKinsey and Co. to advise on the implementation of a coding system for the food industry. From the outset, McKinsey understood that product coding could only succeed on an all-or-nothing basis. Their strategy was thus to take a top-down approach by bringing together the top people from all sections of the grocery industry—retailers, manufacturers, and their trade associations. By bringing all these interests together, it was hoped, a consensus could be reached that would be acceptable to all. Once that consensus had been reached by the people at the top, they would be in a position to guarantee the commitment of their organizations and the rest would follow. In August 1970 an "Ad Hoc Committee of the Grocery Industry" was set up, consisting of "a blue ribbon group of ten top industry leaders, five representing food manufacturers and five representing food distributors," which was chaired by the president of the Heinz Company. The ad hoc committee in turn retained McKinsey as consultants, and in the course of the next few months the consultants made some fifty presentations to the industry. These presentations were invariably given to the top echelon of each organization, and great efforts were made to sell the concept of product coding as a vision of the future.

Simultaneously, many technical decisions had to be made. One of the first was to agree upon a product code of ten digits—five digits representing the manufacturer and five representing the product. This in itself was a compromise between the manufacturers, who wanted a longer code for greater flexibility, and the retailers, who wanted a shorter one to reduce equipment costs. (In fact, this was arguably a bad decision. Europe, which developed its own European Article Numbering system somewhat later, adopted a six-plus-six code.) It was also decided that the product code would be applied by the manufacturer, which would be cheaper than in-store application by retailers; and the article number would contain no price information because the price would differ from store to store.

By mid-1971 the principle of a universal product code had been accepted throughout the industry, but there remained the major issues of agreeing on the technology to be used and pressing the technology into service. This was yet another politically sensitive area. Applying the code to products had to be inexpensive in order not to disadvantage small manufacturers; and the expense of product coding could not add significantly to the cost of goods, as this would disadvantage retailers who were unable to participate in the system because they would be forced to bear the increased cost of the system without deriving any benefit. The checkout equipment also had to be relatively inexpensive because millions of bar-code readers would eventually be needed. These conditions ruled out the use of the expensive magnetic ink and optical character-recognition systems then in use, which had been developed for banks in the 1950s. Various experimental systems were tested, both in the laboratory and in realistic supermarket conditions. These trials cost several million dollars, but at the end of 1971 there was a much better awareness of the complex trade-offs that had to be made—for example, between printing costs and scanner costs. In the spring of 1973 the bar-code system with which we are now all familiar was formally adopted. This system was largely developed by IBM, and it was selected simply for being the cheapest and the most trouble-free option. The scanning technology was inexpensive, the bar code could be printed in ordinary ink, and readers were not overly sensitive to smearing and distortion of the packaging.

By the end of 1973 over 800 manufacturers in the United States had been allocated identification numbers. These included virtually all the largest manufacturers, and between them they accounted for over 84 percent of retail food sales. The first deliveries of bar-coded merchandise began in 1974, along with the availability of scanners manufactured by IBM and NCR.

Because of the enormous reequipment costs involved, the take-up of the new technology was much slower among food retailers than it had been among food manufacturers, who simply had to redesign their labels. But by the mid-1980s most food chains had automated checkouts in their larger stores. It would be some years, however, before the smaller operators could economically adopt bar-code readers.

The bar-code principle diffused very rapidly to cover not just food items but most packaged retail goods (note the ISBN bar code on the back cover of this book). The bar code has become an icon of great economic significance.

Alongside the physical movement of goods there is now a corresponding information flow that enables computers to track the movement of this merchandise. Thus the whole manufacturing-distribution network is becoming increasingly integrated. In retail outlets, shelves do not have to become empty before the restocking process is started. As backroom stock levels fall, orders can be sent automatically and electronically to wholesalers and distribution centers, which in turn can automatically send orders to the manufacturers. Special protocols for the exchange of electronic information among manufacturers, distributors, and retailers in a standard form were developed to expedite this process.

In a sense, there are two systems coexisting, one physical and one virtual. The virtual information system in the computers is a representation of the status of every object in the physical manufacturing and distribution environment—right down to an individual can of peas. This would have been possible without computers, but it would not have been economically feasible because the cost of collecting, managing, and accessing so much information would have been overwhelming. In the old retail establishment, instead of keeping precise inventories in real time, it was in practice often cheaper to overstock items—or even to understock them and risk losing sales.

Although it is true that bar coding caused checkout and inventory costs to be reduced, this was not the primary reason for a renaissance of supermarket profits, which rose to an all-time high in the 1980s. A more important source of profit was a greatly increased product range—in some cases the number of lines more than tripled—as a result of checkout automation. Many of these new lines were luxury goods with large margins; these were the real generators of profits.

For the food industry itself, there were negative impacts of bar coding, much as the rise of the supermarkets in the 1930s spelled an end to the owner-operated grocery store. For example, the barriers to entry into food manufacturing became greater. Not only were there the bureaucratic hurdles of UPC allocation to surmount, but penetrating the retail network of national chains now required at minimum an operation of substantial scale. Indeed, a curious development of the late twentieth century was the emergence of folksy specialty foods—apparently made by small-scale food preparers, but in fact mass-produced and sold by national chains. As I write, I have in front of me a box of the "finest hand-baked biscuits," a gift from a friend returning from a vacation in Devon, England. On the bottom of the carton, out of sight, is a bar code for the supermarket checkout. There could be few more potent ironies of modern living.

In 1820 the English mathematician Charles Babbage invented the Difference Engine, the first fully automatic computing machine. Babbage's friend Ada Lovelace wrote a *Sketch of the Analytical Engine* (1843), which was the best description of the machine until recent times. In the 1970s the programming language ADA was named in Lovelace's honor. BABBAGE AND LOVELACE PORTRAITS COURTESY OF SCIENCE MUSEUM, LONDON; DIFFERENCE ENGINE COURTESY OF IBM, UNITED KINGDOM, LTD.

The Central Telegraph Office, London, routed telegrams between British provincial towns. This 1874 engraving from the *Illustrated London News* shows the buzz of activity as incoming telegrams were received for onward

The Felt & Tarrant Comptometer was designed for rapid addition in offices. The photograph shows these machines in use around the turn of the twentieth century at the Office of the Auditor of Freight Accounts, Pennsylvania Rail Road. COURTESY OF JAMES W. CORTADA

In 1911 Thomas J. Watson Sr. became the general manager of the Tabulating Machine Company. Photograph *(above)* shows Watson addressing a sales school in the early 1920s. In 1924 the firm was renamed International Business Machines (IBM) and became the world's most successful office machine company. Photograph *(below)* shows a typical punched-card office of the 1920s.

Shown with its inventor, Vannevar Bush, in the mid-1930s, the Differential Analyzer was the most powerful analog computing machine developed between the world wars. Bush, a distinguished MIT alumnus, later became chief scientific adviser to President Roosevelt and head of civilian war research during World War II. MIT MUSEUM

The IBM Automatic Sequence Controlled calculator, usually known as the Harvard Mark I, was placed in operation in 1943. Weighing 5 tons and measuring 51 feet across, it could perform several basic operations per second. HARVARD UNIVERSITY

The ENIAC, developed during World War II, was capable of 5,000 basic arithmetic operations per second. It contained 20,000 electronic tubes, consumed 150 kilowatts of electric power, and filled a room 30 x 50 feet. The inventors of the ENIAC, John Presper Eckert and John Mauchly, are seen in the foreground *(left)* and center. UNIVERSITY OF PENNSYLVANIA

John Presper Eckert *(left)* and Herman H. Goldstine holding a decade counter from the ENIAC. This 22-tube unit could store just one decimal digit. SCIENCE MUSEUM, LONDON

During 1944–45 John von Neumann collaborated with the developers of the ENIAC to produce the design for a new computer. The stored-program computer that resulted has been the theoretical basis for almost all computers built since. The photograph shows von Neumann with the computer he built in the early 1950s at the Institute for Advanced Study, Princeton. INSTITUTE FOR ADVANCED STUDY, PRINCETON

The first stored-program computer to operate, in 1948, was the "baby machine" built at Manchester University in England. Although too small to perform realistic calculations, it established the feasibility of the stored-program concept. The photograph shows the developers Tom Kilburn *(left)* and Frederik C. Williams at the control panel. NATIONAL ARCHIVES FOR HISTORY OF COMPUTING, MANCHESTER

Completed at Cambridge University in 1949, the EDSAC was the first full-scale stored-program computer to go into regular service. CAMBRIDGE UNIVERSITY

The UNIVAC, developed by Eckert and Mauchly, was America's first commercially manufactured electronic computer. It attracted publicity on election night 1952, when it was used to forecast the results of the presidential election. HAGLEY MUSEUM AND LIBRARY

System/360, announced in 1964, enabled IBM to dominate the computer industry well into the 1980s. IBM

In 1959 IBM announced the model 1401 computer, a fast and reliable "second-generation" machine that used transistors in place of electronic tubes. IBM

Jay W. Forrester, the leader of Project Whirlwind, with a prototype core memory plane. MIT MUSEUM

SAGE, an early 1960s air defense system, was located in some thirty "direction centers" strategically placed around the country to provide complete radar coverage for the United States against airborne attacks. The consoles were designed for use by ordinary airmen and established the technology of human/computer interaction using screens and light-pens in place of punched cards or the clunky teletypes of previous systems. CHARLES BABBAGE INSTITUTE

In this advertisement, IBM drew attention to its own John Backus, the inventor of the FORTRAN programming language, to convey the depth and competence of its software. Starting in 1969 computer manufacturers began to "unbundle" their software, creating a market for software packages and a competitive threat to the established computer manufacturers. IEEE

In the mid-1960s time-sharing transformed the way people used computers, by providing inexpensive access to a powerful computer with user-friendly software. The time-sharing system developed at Dartmouth College, shown here, exposed most of the college's students to computers using the BASIC programming language. DARTMOUTH COLLEGE LIBRARY

In the mid-1960s the development of microelectronics enabled the building of small, inexpensive computers with the power of a traditional mainframe. The first of these machines was the PDP–8, shown here, tens of thousands of which were sold. COMPUTER MUSEUM, BOSTON

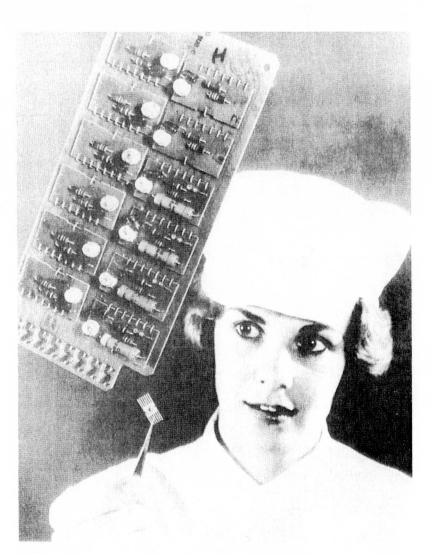

The first computers were constructed using electronic tubes. By the late 1950s this technology had been displaced by discrete transistors, which were smaller, faster, and cheaper, and produced far less heat. In the mid-1960s discrete transistors gave way to integrated circuits, which contained several transistors and other components on a single silicon "chip," and were correspondingly smaller, faster, and cheaper. By 1990 chips were being produced with over a million transistors. During the 1970s the mainstream electronics industry began to appropriate the new digital electronics and integrated circuits, producing a stream of innovative products such as video games, calculators, and digital watches. CHARLES BABBAGE INSTITUTE

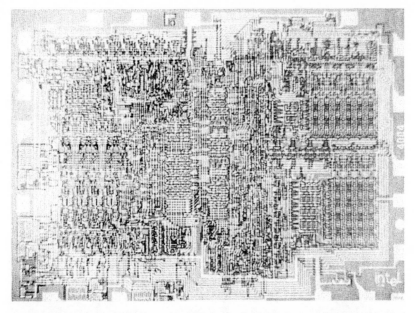

The Intel 4004 Microprocessor, announced in 1971, was the first "computer on a chip." The initial market for microprocessors was expected to be for simple control applications—such as in automobiles, office equipment, and consumer durables. The idea of using microprocessors for full-scale computers emerged gradually, and it was not until 1977 that the first personal computers hit the market. INTEL

The Altair 8800 was the first microprocessor-based computer. It was sold as a kit to electronics hobbyists for $397. Note that the Altair 8800 was termed a "minicomputer"—the term "personal computer" had not yet been coined. ROBERT VOELKER

The Apple II, launched in 1977, established the paradigm of the personal computer: a central processing unit equipped with a keyboard and screen, and floppy disk drives for program and data storage. It was successful both as a consumer product and a business machine. CHARLES BABBAGE INSTITUTE

In 1981 the corporate personal computer was fully legitimated by the announcement of the IBM PC, which rapidly became the industry standard. Rather than make a "clone" of the IBM PC, Apple Computer decided to produce the Macintosh, shown here, a low-cost computer with user-friendly "point-and-click" software. The Macintosh was launched in 1984, and its user-friendliness was unrivaled for several years. CHARLES BABBAGE INSTITUTE

Bill Gates has engendered more interest and controversy than any other figure in the history of the
computer. The richest man in America, Gates personally owns approximately a third of Microsoft,
which developed the MS-DOS and Windows operating systems installed on hundreds of millions of
personal computers. Gates, who is simultaneously
portrayed as an archetypal computer nerd and a
ruthless, Rockefeller-style businessman, has come
to epitomize the entrepreneurs riding the roller-
coaster of the *fin-de-siècle* information economy.
MICROSOFT

During the 1960s and 1970s, J. C. R. Licklider set
in place many of the research programs that cul-
minated in the user-friendly personal computer
and the Internet. A psychologist by training, Licklider was a consummate political operator, who
motivated a generation of computer scientists and
obtained government funding for them to work in
the fields of human/computer interaction and networked computing. MIT MUSEUM

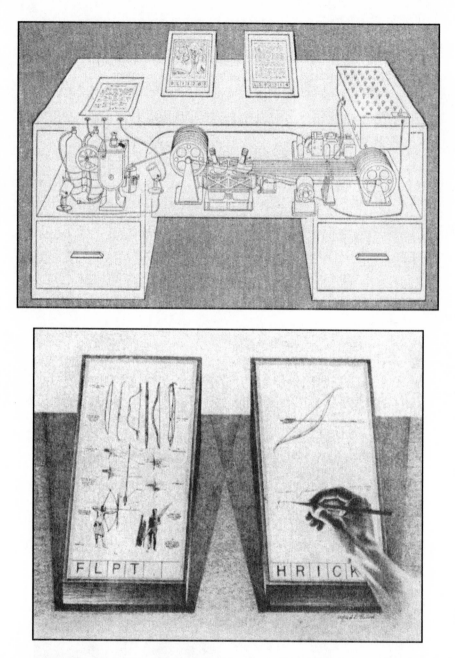

In 1945 Vannevar Bush proposed the "memex," an information machine to help deal with the post-war information explosion. His concept was an important influence on hypertext and other innovations that eventually led to the World Wide Web. The upper illustration shows the microfilm mechanism which would store vast amounts of information; the lower illustration shows the viewing screens close up. COURTESY TIME INC.

8

:: SOFTWARE

OF ALL the elements that make up the technology of computing, none has been more problematic or more unpredictable than software. When computers first arrived on the scene many people thought that, say, artificial intelligence would be difficult and would perhaps take half a century to realize; but no one ever expected that software would take a similar time to mature. The fundamental difficulty of writing software was that, until computers arrived, human beings had never before had to prepare detailed instructions for an automaton—a machine that obeyed unerringly the commands given to it, and for which every possible outcome had to be anticipated by the programmer.

Historically, the nearest artifact to a modern computer program was a mathematical "algorithm." The idea of an algorithm—a procedure for doing something—is very ancient, dating from at least the Babylonians. A familiar example of an algorithm in recreational mathematics is a procedure to generate a magic square. A magic square is a grid consisting of n rows of n cells, in which the integers from 1 to n^2 are placed so that each row, column, and diagonal adds up to the same value—for example, 65 in the case of the order-5 magic square below.

17	24	1	8	15
23	5	7	14	16
4	6	13	20	22
10	12	19	21	3
11	18	25	2	9

A typical algorithm to produce a magic square with an odd number of rows and columns was given by the English mathematician Charles Hutton in an 1814 book titled *Recreations in Mathematics and Philosophy,* which has the delightful subtitle *Amusing Dissertations and Enquiries Concerning a Variety of Subjects Most Remarkable and Proper to Excite Curiosity and Attention to the Whole Range of the Mathematical and Philosophical Sciences.* Here is Hutton's algorithm (of which the flavor is more important than the detail):

> We shall here suppose an odd square, the root of which is 5, and that it is required to fill up with the first 25 of the natural numbers. In this case, begin by placing unity in the middle cell of the horizontal band at the top; then proceed from left to right, ascending diagonally, and when you go beyond the square, transport the next number 2 to the lowest cell of that vertical band to which it belongs; set 3 in the next cell, ascending diagonally from left to right, and as 4 would go beyond the square, transport it to the most distant cell of the horizontal band to which it belongs; set 5 in the next cell, ascending diagonally from left to right, and as the following cell, where 6 would fall, is already occupied by 1, place 6 immediately below 5; place 7 and 8 in the two next cells, ascending diagonally, as seen in the figure; and then, in consequence of the first rule of transposition, set 9 at the bottom of the last vertical band; then 10, in consequence of the second, in the last cell on the left of the second horizontal band; then 11 below it, according to the third rule: after which continue to fill up the diagonal with the numbers 12, 13, 14, 15; and as you can ascend no farther, place the following number 16 below 15; if you then proceed in the same manner, the remaining cells of the square may be filled up without any difficulty, as seen in the above figure.

The point about this algorithm is that although it is peppered with ambiguities and the method is explained largely by means of an example, any reasonably intelligent human being would have no difficulty in generalizing it and, in principle, could produce a magic square of any size. However, if called upon to make a large magic square (of side greater than 15, say), the human computer would almost certainly make an arithmetic error along the way.

A digital computer, on the other hand, has almost the opposite characteristics of a human computer. There is no presently available digital computer and software that could take Hutton's algorithm as written and use it to produce a magic square. Before a digital computer could make use of Hutton's algorithm, a human programmer would first have to recast it as a computer program that resolved all the ambiguities. But once that had been done correctly, the digital computer could produce magic squares forever without any possibility of error, however large the square. Converting this algorithm into a computer program is harder than it looks, but its logical complexity is almost trivial compared with that of a large modern computer program: Imagine how much simpler it is, for example, than an airline reservations system. However, the magic squares problem was just about at the level of complexity that could be handled by the very first prototype computers in the late 1940s and early 1950s, where the roots of modern software lie.

A Subroutine Library

The first operational stored-program computer was the Cambridge EDSAC, which sprang to life in May 1949, and therefore the Cambridge group was the first to develop early ideas about computer programming. By 1948 construction of the EDSAC was well under way and the project director, Maurice Wilkes, began to turn his thoughts away from hardware and toward the programming problem. Inside every stored-program computer (including those of today) programs are stored in pure binary numbers—a pattern of machine representations of ones and zeros. For example, in the EDSAC, the single instruction "Add the short number in memory location 25" was stored as:

11100000000110010

As a programming notation, binary code was unacceptable because humans cannot readily remember long strings of binary digits. It was therefore an obvious idea that occurred to all the computer groups to use a more programmer-friendly notation when designing and writing programs. In the case of the EDSAC, the instruction "Add the short number in memory location 25" was written

A 25 S

where A stood for "add," 25 was the decimal address of the memory location, and S indicated that a "short" number was to be used.

Although von Neumann and Goldstine had come up with a similar programming notation for the EDVAC, they had always assumed that the conversion of the human-readable program into binary code would be done by the programmer—or by a more lowly technician "coder." Wilkes, however, realized that this conversion could be done by the computer itself. After all, a computer was designed to manipulate numbers, and letters of the alphabet were represented by numbers inside a computer; and in the last analysis, a program was nothing more than a set of letters and numbers. This was an insight that is obvious only in retrospect.

In October 1948 the Cambridge group was joined by a twenty-one-year-old research student, David Wheeler, who had just been awarded a Cambridge "first" in mathematics. Wilkes turned over the programming problem to Wheeler as the major part of his Ph.D. topic. To translate programs into binary, Wheeler wrote a very small program, which he called the Initial Orders, that would read in a program that was punched in symbolic form on teleprinter tape, convert it to binary, and place it in the memory of the computer ready for execution. This tiny program consisted of just thirty instructions that he had honed to perfection. In many ways programming, like mathematics, was a young person's game. Writing a program called for a similar kind of fresh outlook, unbound by convention, and a very high order of manipulative technique. The programs that Wheeler and others wrote at Cambridge during 1948–51 have something of the same appeal as elegant theorems in mathematics.

But as the Initial Orders solved one problem—converting a symbolic program into binary—another problem came into view: actually getting programs to work correctly. A program consisted of a hundred or perhaps a thousand instructions, each of which had to be exactly right. A single error in any one of the instructions would cause the whole program to fail—either producing incorrect results or simply grinding to a halt. Before people began to write real programs for real computers, it had always been assumed that there would be no particular difficulty in getting programs to work. Consequently, there was a surprise in store for whichever group was the first to get a computer running. This of course was the Cambridge EDSAC, and the problems surfaced within a few weeks of the machine first operating. Wilkes later recalled:

> By June 1949 people had begun to realize that it was not so easy to get a program right as had at one time appeared. I well remember when this realization first came on me with full force. The EDSAC was on the top floor of the building and the tape-punching and editing equipment one floor below on a gallery that ran round the room in which the differential analyzer was installed. I was trying to get working my first non-trivial program, which was one for the numerical integration of Airy's differential equation. It was on one of my journeys between the EDSAC room and the punching equipment

that "hesitating at the angles of stairs" the realization came over me with full force that a good part of the remainder of my life was going to be spent in finding errors in my own programs.

At first, mistakes in programs were simply called mistakes, or errors. But within a few years they were called "bugs" and the process of correcting them was called "debugging."

Quite a lot could be done to reduce the incidence of errors by checking a program very carefully before the programmer attempted to run it. It was found that half an hour spent checking a program would save lots of time later trying to fix it—not to mention the machine time which, in the early 1950s, cost perhaps $50 an hour. Even so, it became almost a tradition that newcomers to programming simply could not be told—they almost all had to learn the truth about debugging the same painfully slow and expensive way. The same remains true today.

The Cambridge group decided that the best way of reducing errors in programs would be to develop a "subroutine library"—an idea that had already been proposed by von Neumann and Goldstine. It was realized that many operations were common to different programs—for example, calculating a square root or a trigonometrical function, or printing out a number in a particular format. The idea of a subroutine library was to write these operations as kind of miniprograms that programmers could then copy into their own programs. In this way a typical program would consist of perhaps two-thirds subroutines and one-third new code. (At this time a program was often called a routine— hence the term *subroutine*. The term *routine* died out in favor of *program* in the 1950s, but the term *subroutine* has stayed with us.) The idea of reusing existing code was and remains the single most important way of improving programmer productivity and program reliability.

Naturally, Wilkes gave David Wheeler the job of organizing the details of the subroutine library. There was much more to this than met the eye—one has only to look at the number of different schemes developed in the 1950s before these ideas became standardized. Placing the subroutines in the computer memory was rather like packing a suitcase. Some people were inclined to throw in everything they needed and then move things around if the lid would not shut. Other people divided the suitcase into equal-sized compartments—which was easy to manage but wasteful, because a compartment big enough for a shirt was much larger than was necessary for a pair of socks. Other groups treated the magnetic-tape and drum stores as an enormous trunk, and fished out a fresh copy of a subroutine every time it was needed. Wheeler's solution was simple and elegant. Each subroutine was written so that, when it was loaded into the memory immediately following the previous subroutine in the program, it would run correctly. Thus all the components of a program were loaded into the bottom of the store working upward without any gaps, making the best possible use of every one of the precious few hundred words of memory.

In 1951 the Cambridge group put all their programming ideas into a textbook entitled *The Preparation of Programs for an Electronic Digital Computer*, which Wilkes chose to publish in the United States in order to reach the widest possible readership. At this date the first research computers were only just coming into operation, and generally the designers had given precious little thought to programming until they suddenly had a working machine. The Cambridge book was the only programming textbook then available, and it rushed in to fill the vacuum. The Cambridge model thus set the programming style for the early 1950s, and even today the organization of subroutines in virtually all computers still follows the Cambridge model.

Programming Languages: FORTRAN and COBOL

By about 1953, however, the center of programming research had moved from England to the United States, where there was heavy investing in computers with large memories, backed up with magnetic tapes and drums. Computers now had perhaps ten times the memory of the first prototypes. The programming systems designed to be shoehorned into the first computers with tiny 1,000-word memories were no longer appropriate, and there was a desire to exploit the increased potential of "large" memories (as much as 10,000 words) by designing more sophisticated programming techniques. Moreover, the patient crafting of programs, instruction by instruction, was becoming uneconomic. Programs cost too much in programmer time and took too long to debug.

A number of American research laboratories and computer manufacturers began to experiment with "automatic programming" systems that would enable the programmer to write a program in some high-level programming code, which was easy for humans to write since it looked like English or algebra, but which the computer could convert into binary machine instructions. Automatic programming was partly a technological problem and partly a cultural one. In many ways it was the latter that was more difficult. There is no craft that is so tradition-bound as one that is five years old, and many of the seasoned programmers of the 1950s were as reluctant to change their working methods as had been the hand weavers to adopt looms two hundred years earlier.

Probably no one did more to change the conservative culture of 1950s programmers than Grace Hopper, the first programmer for the Harvard Mark I in 1943, then with UNIVAC. For several years in the 1950s she barnstormed around the country, proselytizing the virtues of automatic programming at a time when the technology delivered a good deal less than it promised. At UNIVAC, for example, Hopper had developed an automatic programming system she called the A-0 compiler. (Hopper chose the word *compiler* because the system would automatically put together the pieces of code that made up the complete program.) The A-0 compiler worked, but the results were far too inefficient to be commercially acceptable. Even quite simple programs would take as long as an hour to translate, and the resulting programs were woefully slow.

Thus, in 1953–54 the outstanding problem in programming technology was to develop an automatic programming system that produced programs as good as those written by an experienced programmer. The most successful attack on this problem came from IBM, in a project led by a twenty-nine-year-old researcher named John Backus. The system he produced was called the Formula Translator—or FORTRAN for short. FORTRAN became the first really successful "programming language," and even today it remains the lingua franca of scientific computing. In FORTRAN a single statement would produce many machine instructions, thus greatly magnifying the expressive power of a programmer. For example, the single statement

$$X = 2.0 * FSIN(A + B)$$

would produce all the code necessary to calculate the function $2\sin(a + b)$ and assign the value to the variable denoted by x.

Backus made the case for FORTRAN mainly on economic grounds. It was estimated in 1953 that half the cost of running a computer center was accounted for by the salaries of programmers, and that from one-quarter to one-half of the average computer's available time was spent in program testing and debugging. Backus argued that "programming and debugging accounted for as much as three-quarters of the cost of operating a computer; and obviously, as computers got cheaper, this situation would get worse." This was the cost-benefit analysis that Backus made to his boss at IBM in late 1953 in a successful bid to get a budget and staff to develop FORTRAN for IBM's soon-to-be-announced model 704 computer. During the first half of 1954, Backus began to put together his programming team.

The FORTRAN project did not have a high profile at IBM—as evidenced by its unglamorous location in an annex to IBM's headquarters on Madison Avenue, "on the nineteenth floor . . . next to the elevator machinery." FORTRAN was regarded simply as a research project with no certain outcome, and it had no serious place in the product plan for the 704 computer. Indeed, it was not at all clear in 1954 that it would even be possible to make an automatic programming system that would produce code as good as that written by a human programmer. There were plenty of skeptics in IBM who thought Backus was attempting the impossible. Thus, producing code as good as a human being was always the overriding objective of the FORTRAN team. In 1954 the group produced a preliminary specification for the language. Very little attention was paid to the elegance of the language design because everything was secondary to efficiency. It never occurred to Backus and his colleagues that they were designing a programming language that would still be around in the twenty-first century.

Backus and his team now began to sell the idea of FORTRAN to potential users. Most of the feedback they received was highly skeptical, coming from seasoned programmers who had already become cynical about the supposed virtues of automatic programming: "[T]hey had heard too many glowing descriptions of what turned out to be clumsy systems to take us seriously." Work began on writing the translator early in 1955 and Backus planned to have the project completed in six months. In fact it took a team of a dozen programmers two and a half years to write the 18,000 instructions in the FORTRAN system. The FORTRAN translator was by no means the longest program that had been written up to that date, but it was a big program and very complex. Much of the complexity was due to the attempt to get the compiler to produce programs that were as good as those written by human programmers. The schedules continued to slip, and during 1956 and 1957 the pace of debugging was intense. To get as much scarce machine time as possible the team would make use of the slack night shift—"often we would rent rooms in the Langdon Hotel . . . on 56th Street, sleep there a little during the day and then stay up all night."

The first programmer's manual for FORTRAN, handsomely printed between glossy IBM covers, announced that FORTRAN would be available in October 1956, which turned out to be "a euphemism for April 1957"—the date on which the system was finally released.

One of the first users of the FORTRAN system was the Westinghouse-Bettis nuclear power facility in Maryland. Late one Friday afternoon, 20 April 1957, a large box of punched cards arrived at the computer center. There was no identification on the cards, but it was decided that they could only be for the "late 1956" FORTRAN compiler. The programmers decided to try to run a small test program. They were amazed when the FORTRAN system swallowed the test program and produced a startlingly explicit diagnostic statement on the printer: "SOURCE PROGRAM ERROR . . . THE RIGHT PARENTHESIS IS NOT FOLLOWED BY A COMMA." The error was fixed and the program re-translated; it then proceeded to print out correct results for a total of twenty-two minutes. "Flying blind," the Westinghouse-Bettis computer center had become the first FORTRAN user.

In practice the FORTRAN system produced programs that were 90 percent as good as those written by hand, as measured by the memory they occupied or the time they took to run. It was a phenomenal aid to the productivity of a pro-grammer. Programs that had taken days or weeks to write and get working could now be completed in hours or days. By April 1958, half of the twenty-six IBM 704 installations surveyed by IBM were using FORTRAN for half of their problems. By that fall, some sixty installations were using the system. Mean-while, Backus's programming team was rapidly responding to the feedback from users by preparing a new version of the language and compiler called FORTRAN II. This contained 50,000 lines of code and took fifty man-years to develop before it was released in 1959.

Although IBM's FORTRAN was the most successful scientific programming language, there were many parallel developments by other computer manufac-turers and computer users around the same time. For example, Hopper's group at UNIVAC had improved the A-0 compiler and supplied it to customers as the Math-Matic programming language. Within the academic community, pro-gramming languages such as MAD at the University of Michigan and IT at Carnegie-Mellon were developed. There was also an attempt to develop an in-ternational language, called ALGOL, by the Association for Computing Ma-chinery in the United States and its European sister organizations. Several dozen different scientific programming languages were in use by the late 1950s.

However, within a couple of years FORTRAN had come to dominate all the other languages for scientific applications. It has sometimes been supposed that this was the result of a conscious plan by IBM to dominate the computer scene, but there is no real evidence for this. What actually happened was that FORTRAN was the first effective high-level language to achieve widespread use. In 1961 Daniel D. McCracken published the first FORTRAN textbook, and universities and colleges began to use it in their undergraduate program-ming courses. In industry, standardizing on FORTRAN had the advantage of making it possible to exchange programs with other organizations that used different makes of computer—FORTRAN would work on any machine for which a translator was available. Also there was the advantage of a growing supply of FORTRAN-trained programmers on the job market. Soon, virtually all scientific computing in the United States was done in FORTRAN; while ALGOL survived for a few years in Europe, it was soon ousted by FORTRAN there too. Always, the overwhelming attraction of FORTRAN was that it had become a "standard" language accepted by the whole user community. In 1966 FORTRAN became the first programming language to be formally standardized by the American National Standards Institute.

While FORTRAN was becoming the standard for scientific applications, another language was emerging for business applications. FORTRAN had become a standard language largely by accident, COBOL (an acronym for COmmon Business-Oriented Language) in effect was created as a standard by the action of the U.S. government. By the time FORTRAN was accepted as a standard, there had been a serious economic loss to industry and government because, whenever they changed their computers, they had to rewrite all their programs. This was expensive and often disruptive. The government was determined not to let this happen with its commercial application programs, so in 1959 it sponsored a Committee on Data Systems and Languages (CODASYL) to design a new standard language for commercial data processing. This language emerged as COBOL 60 the following year. By writing their programs in the new language, users would be able to retain all their existing software investment when they changed computers.

Grace Hopper played an active role in promoting COBOL. Although she was not one of the language designers, it was heavily influenced by the Flow-Matic data-processing language she had developed at UNIVAC. COBOL was particularly influenced by what one critic described as Hopper's "missionary zeal for the cause of English language coding." In COBOL a statement designed to compute an employee's net pay by deducting tax from the gross pay might have been written:

SUBTRACT TAX FROM GROSS PAY GIVING NET PAY.

This provided a comforting illusion for administrators and managers that they could read and understand the programs their employees were writing, even if they could not write them themselves. While many technical people could see no point in such "syntactic sugar," it had important cultural value in gaining the confidence of decision makers.

Manufacturers were initially reluctant to adopt the COBOL standard, preferring to differentiate their commercial languages from their competitors—thus IBM had its Commercial Translator, Honeywell had FACT, and UNIVAC had Flow-Matic. Then, in late 1960 the government declared that it would not lease or purchase any new computer without a COBOL compiler unless its manufacturer could demonstrate that its performance would not be enhanced by the availability of COBOL. No manufacturer ever attempted such a proof, and they all immediately fell in line and produced COBOL compilers.

Although for the next twenty years COBOL and FORTRAN would dominate the world of programming languages—between them accounting for 90 percent of applications programs—designing new programming languages for specific application niches remained the single biggest software research activity in the early 1960s. Hundreds of languages were eventually produced, most of which have died out.

Software Contractors

Although programming languages and other utilities helped computer users to develop their software more effectively, programming costs remained the single largest expense of running a computer installation. Many users preferred to minimize this cost by obtaining ready-made programs wherever possible, instead of writing their own.

Recognizing this preference, in the late 1950s computer manufacturers began to develop applications programs for specific industries such as insurance, banking, retailing, and manufacturing. Programs were also produced for applications that were common to many industries, such as payroll, cost accounting, and stock control. All these programs were supplied at no additional charge by the computer manufacturer. There was no concept in the 1950s of software being a salable item: Most manufacturers saw applications programs simply as a way of selling hardware. Indeed, applications programming was often located within the marketing division of the computer manufacturers. Of course, although the programs were "free," the cost of developing them was ultimately reflected in the total price of the computer system.

Another source of software for computer users was the free exchange of programs within cooperative user groups such as SHARE for users of IBM computers and USE for UNIVAC users. At these user groups, programmers would trade programming tips and programs. IBM also facilitated the exchange of software between computer users by maintaining a library of customer-developed programs. These programs could rarely be used directly as they were supplied, but a computer user's own programmers could often adapt an existing program more quickly than writing one from scratch.

The high salaries of programmers, and the difficulty of recruiting and managing them, discouraged many computer users from developing their own software. This created the opportunity for the new industry of *software contracting*: firms that wrote programs to order for computer users. In the mid-1950s there were two distinct markets for the newly emerging software contractors. One market, which came primarily from government and large corporations, was for very large programs that the organizations did not have the technological capability to develop for themselves. The second market was for "custom" programs of modest size, for ordinary computer users who did not have the in-house capability or resources to develop software.

The first major firm in the large-systems sector of software contracting was the RAND Corporation, a government-owned defense contractor. It developed the software for the SAGE air-defense project. When the SAGE project was begun in the early 1950s, although IBM was awarded the contract to develop and manufacture the mainframe computers, writing the programs for the system—estimated at a million lines of code—was way beyond IBM's or anyone else's experience. The contract for the SAGE software was therefore given to the RAND Corporation in 1955. Although lacking any actual large-scale software writing capability, the corporation was judged to have the best potential for developing it. To undertake the mammoth programming task, the corporation created a separate organization in 1956, the System Development Corporation (SDC), which became America's first major software contractor.

The SAGE software development program has become one of the great episodes in software history. At a time when the total number of programmers in the United States was estimated at about 1,200 people, SDC employed a total staff of 2,100, including 700 programmers, on the SAGE project. The central operating program amounted to nearly a quarter of a million instructions, and ancillary software took the total to over a million lines of code. It has been stated that the SAGE software development was a university for programmers. It was an effective, if unplanned, way to lay the foundations of the supremacy of the U.S. software industry.

In the late 1950s and early 1960s several other major defense contractors and aerospace companies—such as TRW, the MITRE Corporation, and Hughes Dynamics—entered the large-scale software contracting business. The only other organizations with the technical capability to develop large software systems were the computer manufacturers themselves. By developing large one-of a-kind application programs for customers, IBM and the other mainframe manufacturers became (and remain) an important sector of the software contracting industry. IBM, for example, having collaborated with American Airlines to develop the SABRE airline reservations system, proceeded to develop reservations systems for other carriers including Delta, Pan American, Eastern, and Alitalia and BOAC in Europe. Most of the other mainframe computer manufacturers also became involved in software contracting in the 1960s, usually building on the capabilities inherited from their office-machine pasts: For example, NCR developed retailing applications, and Burroughs targeted the banks.

While the major software contractors and mainframe computer manufacturers developed very large systems for major organizations, there remained a second market of developing software for medium-sized customers. The big firms did not have the small-scale economies to tap this market effectively, and it was this market opportunity that the first programming entrepreneurs exploited. Probably the first small-time software contractor was the Computer Usage Company, whose trajectory was typical of the sector.

The Computer Usage Company (CUC) was founded in New York City in March 1955 by two scientific programmers from IBM. It was not expensive to set up in the software contracting business, because "all one needed was a coding pad and a pencil," and it was possible to avoid the cost of a computer, either by renting time from a service bureau or by using a client's own machine. CUC was established with $40,000 capital, partly private money and partly secured loans. The money was used to pay the salaries of a secretary and four female programmers, who worked from one of the founders' private apartments until money started coming in. The founders of CUC came from the scientific programming division of IBM, and most of their initial contracts were with the oil, nuclear power, and similar industries; in the late 1950s the company secured a number of NASA contracts. By 1959 CUC had built up to a total of fifty-nine employees. The company went public the following year, which netted $186,000. This money was used to buy the firm's first computer.

By the early 1960s CUC had been joined by several other start-up software contractors. These included the Computer Sciences Corporation (co-founded by a participant of IBM's FORTRAN project team), the Planning Research Corporation, Informatics, and Applied Data Research. These were all entrepreneurial firms with similar growth patterns to the Computer Usage Company, although with varied application domains. Another major start-up was the University Computing Company in 1965.

The first half of the 1960s was a boom period for software contractors. By this date the speed and size of computers, as well as the machine population, had grown by at least an order of magnitude since the 1950s. This created a software-hungry environment in which private and government computer users were increasingly contracting out big software projects. In the public sector the defense agencies were sponsoring huge data-processing projects, while in the private sector banks were moving into real-time computing with the use of automatic teller machines.

By 1965 there were between forty and fifty major software contractors in the United States, of whom a handful employed more than a hundred programmers and had annual sales in the range of $10 million to $100 million. CUC, for example, had grown to become a major firm, diversifying into training, computer services, facilities management, packaged software, and consultancy, in addition to its core business of contract programming. By 1967 it had 12 offices, 700 staff members, and annual sales of $13 million.

The major software contractors, however, were just the tip of an iceberg beneath which lay a very large number of small software contractors, typically with just a few programmers. One estimate in 1967 stated that there were 2,800 software contracting firms in the United States.

The specialist software contractors had also been joined by computer services and consulting firms such as Ross Perot's EDS, founded in 1962, and Management Science America (MSA), formed in 1963, which were finding contract programming an increasingly lucrative business opportunity.

The Software Crisis

By the early 1960s the programming problem was apparently well under control. The term *software* had come into use during 1959–60, and this in itself was an implicit recognition that software was viewed as an entity in its own right, complementary to the hardware. Applications programs that had once seemed formidably difficult had become relatively straightforward using the new technology of programming languages. Within the computer industry, writing-language translators and other operating software had moved from being a research topic to becoming a standard practice. And computer users who did not have a software writing capability could obtain custom programs to order from a software contractor.

Nevertheless, within five years the computer world was mired in what came to be called the "software crisis." The cause was that the power and size of computers were growing much faster than the capability of software designers to exploit them. In the five years between 1960 and the arrival of System/360 and other third-generation computers, the memory size and speed of computers both increased by a factor of ten—giving an effective performance improvement of a hundred. During the same period, software technology had advanced hardly at all. The programming technology that had been developed up to this time was capable of writing programs with 10,000 lines of code, but had problems dealing with programs that were ten times longer. Projects involving a million lines of code frequently ended in disaster. The 1960s was the decade of the software debacle; and the biggest debacle of all was IBM's operating system OS/360.

The "operating system" was the raft of supporting software that all the mainframe manufacturers were supplying with their computers by the late 1950s. It included all the software, apart from programming languages, that the programmer needed to develop applications programs and run them on the computer. One component of the operating system, for example, was the input/output subroutines, which enabled the user to organize files of data on magnetic tapes and disk drives. The programming of these peripheral devices at the primitive physical level was very complex—involving programmer-years of effort—but the operating system enabled the user to deal not with physical items, such as the specific location of bits of data on a magnetic disk, but with logical data, such as a file of employee records. Another operating system component was known as a "monitor" or "supervisor," which organized the flow of work through the computer. For IBM's second-generation computers, the operating software consisted of about 30,000 lines of code. By the early 1960s a powerful operating system was a competitive necessity for all computer manufacturers.

When IBM began planning System/360 in 1962, software had become an essential complement to the hardware, and it was recognized that the software development effort would be huge—an initial estimate put the development budget at $125 million. No less than four distinct operating systems were planned—later known as BOS, TOS, DOS, and OS/360. BOS was the Batch Operating System for small computers; TOS was the operating system for medium-sized magnetic-tape-based computers; and DOS was the disk operating system for medium and large computers. These three operating systems represented the state of the art in software technology and were all completed within their expected time scales and resources. The fourth operating system, OS/360, was much more ambitious, however, and became the most celebrated software disaster story of the 1960s.

The leader of the OS/360 project was one of Howard Aiken's most able Ph.D. students, a software designer in his mid-thirties named Frederick P. Brooks Jr., who subsequently left IBM to form the computer science department at the University of North Carolina. An articulate and creative individual, Brooks was to become a leading advocate in the 1970s of the push to develop an engineering discipline for software construction, and the author of the most famous book on software engineering, *The Mythical Man-Month*.

Why was OS/360 so difficult to develop? The essential problem was that OS/360 was the biggest and most complex program artifact that had ever been attempted. It would consist of hundreds of program components, totaling more than a million lines of code, all of which had to work in concert without a hiccup. Brooks and his co-designers had reasonably supposed that the way to build a very large program was to employ a large number of programmers. This was rather like an Egyptian pharaoh wanting to build a very large pyramid by using a large number of slaves and a large number of stone blocks. Unfortunately, this only works with simple structures; software does not scale up in the same way, and this fact lay at the heart of the software crisis.

An additional difficulty was that OS/360 was to exploit the new technology of "multiprogramming," which would enable the computer to run several programs simultaneously. The risk of incorporating this unproved technology was fully appreciated at the time and, indeed, there was much internal debate about it. But multiprogramming was a marketing necessity, and at the System/360 launch on 7 April 1964 OS/360 was "the centerpiece of the programming support," with delivery of a full multiprogramming system scheduled for mid-1966.

The development of the OS/360 control program—the heart of the operating system—was based at the IBM Program Development Laboratories in Poughkeepsie, New York. There it had to compete with other System/360 software projects that were all clamoring for the company's best programmers. The development task got under way in the spring of 1964 and was methodically organized from the start—with a team of a dozen program designers leading a team of sixty programmers implementing some forty functional segments of code. But like almost every other large software project of the period, the schedules soon began to slip. Not for any specific reason, but for a myriad of tiny causes. As Brooks explained:

> Yesterday a key man was sick, and a meeting couldn't be held. Today the machines are all down, because lightning struck the building's power transformer. Tomorrow the disk routines won't start testing, because the first disk

is a week late from the factory. Snow, jury duty, family problems, emergency meetings with customers, executive audits—the list goes on and on. Each one only postpones some activity by a half-day or a day. And the schedule slips, one day at a time.

More people were added to the development team. By October 1965 some 150 programmers were at work on the control program, but a "realistic" estimate now put slippage at about six months. Early test trials of OS/360 showed that the system was "painfully sluggish" and that the software needed extensive rewriting to make it usable. Moreover, by the end of 1965 fundamental design flaws emerged for which there appeared to be no easy remedy. For the first time, OS/360 "was in trouble from the standpoint of sheer technological feasibility."

In April 1966 IBM publicly announced the rescheduling of the multiprogramming version of OS/360 for delivery in the second quarter of 1967—some nine months later than originally announced. IBM's OS/360 software problems were now public knowledge. At a conference of anxious IBM users, the company's chairman, Tom Watson Jr., decided the best tactic was to make light of the problems:

> A few months ago IBM's software budget for 1966 was going to be forty million dollars. I asked Vin Learson [the head of the System/360 development program] last night before I left what he thought it would be, and he said, "Fifty million." This afternoon I met Watts Humphrey, who is in charge of programming production, in the hall here and said, "Is this figure about right? Can I use it?" He said, "It's going to be sixty million." You can see that if I keep asking questions we won't pay a dividend this year.

But beneath Watson's levity, inside IBM there was a "growing mood of desperation." The only way forward appeared to be to assign yet more programmers to the task. This was later recognized by Brooks as being precisely the wrong thing to do: "Like dousing a fire with gasoline, this makes matters worse, much worse. More fire requires more gasoline, and thus begins a regenerative cycle which ends in disaster." Writing a major piece of software was a subtle creative task and it did not help to keep adding more and more programmers: "The bearing of a child takes nine months, no matter how many women are assigned."

Throughout 1966 more and more people were added to the project. At the peak more than 1,000 people at Poughkeepsie were working on OS/360—programmers, technical writers, analysts, secretaries, and assistants—and all together some 5,000 staff-years went into the design, construction, and documentation of OS/360 between 1963 and 1966.

OS/360 finally limped into the outside world in mid-1967, a full year late. By the time OS/360 and the rest of the software for System/360 had been delivered to customers, IBM had spent half a billion dollars on it—four times the original budget—which, Tom Watson explained, made it "the single largest cost in the System/360 program, and the single largest expenditure in company history." But there was perhaps an even bigger, human cost:

> The cost to IBM of the System/360 programming support is, in fact, best reckoned in terms of the toll it took of people: the managers who struggled to make and keep commitments to top management and to customers, and the programmers who worked long hours over a period of years, against obstacles of every sort, to deliver working programs of unprecedented complexity. Many in both groups left, victims of a variety of stresses ranging from technological to physical.

Had OS/360 merely been late, it would not have become such a celebrated disaster story. But when it was released, the software contained dozens of errors that took years to eradicate. Fixing one bug would often simply replace it with another. It was like trying to fix a leaking radiator. The unreliability of big programs lay at the heart of the software crisis.

Software Engineering

To the general public, the best-known catastrophe due to a software error occurred with the Mariner I spacecraft, the first U.S. space vehicle designed to visit another planet (Venus). On the morning of 22 July 1962, the space vehicle rocketed from the launch pad and four minutes into its flight began moving on an erratic path. It had to be destroyed in the air before it could do any damage. Subsequent investigation showed that the fault was caused by a single incorrect character in the equations of motion encoded in a guidance program. The Mariner I bug became emblematic of the software crisis. The story was embellished by programming gurus, retold in computer textbooks, and rehashed in the popular press. It required little imagination to see the terrible consequences if a similar software problem were to occur in a manned space flight, a nuclear power plant, or a fly-by-wire airplane. It began to look as though, however much computers advanced, in practice it would not be possible to develop programs that could use their power.

During 1967 there was a groundswell of frustration and anxiety in the software writing community—frustration because software people were unable to exploit the potential of the rapidly advancing hardware and anxiety because they were experiencing the personal trauma of managing projects that were going disastrously wrong. A group of academics and industrial software developers decided to organize a worldwide working conference. This conference, which was held in Garmisch-Partenkirchen, Germany, in October 1968, was entitled Software Engineering, a term coined by the organizers: "The phrase 'software engineering' was deliberately chosen as being provocative, in implying the need for software manufacture to be based on the types of theoretical foundations and practical disciplines that are traditional in the established branches of engineering."

Software development had a long way to go before it could become a true engineering discipline. One participant from Bell Labs, which was then embroiled in developing an operating system called Multics with MIT and General Electric, compared software writers with hardware developers: "they are the industrialists and we are the crofters. Software production today appears in the scale of industrialization somewhere below the more backward construction industries." Although there were some serious academic and industrial papers presented at the conference, the real importance of the meeting was to act as a kind of encounter group for senior figures in the world of software to trade war stories. One participant from MIT confessed: "We build systems like the Wright brothers built airplanes—build the whole thing, push it off the cliff, let it crash, and start over again."

The Garmisch conference began a major cultural shift in the perception of programming. Software writing started to make the transition from being a craft for a long-haired programming priesthood to becoming a real engineering discipline. It was the transformation from an art to a science: As in constructing a bridge, it was important that a software artifact should have an aesthetic appeal, but it was far more important that it should not fall down.

The search began for widely applicable tools and working practices that could be used to engineer better software. These ideas included structured design, formal methods, and development models. All of these were aimed at managing the inherent complexity of writing large programs.

The most widely adopted engineering practice was the use of a "structured design methodology." This took the view that the best way to manage complexity was to limit the software writer's field of view. Thus, the programmer would begin with an overview of the software artifact to be built. When it was satisfactory it would be put to one side, and all the programmer's attention would be directed to a lower level of detail—and so on, until eventually actual program code would be written at the lowest level in the design process. Structured programming was a good intuitive engineering discipline and was probably the most successful programming idea of the 1970s. A whole subculture of consultants and gurus peddled the patent medicine of structured programming, and techniques for structured programming in FORTRAN and COBOL were developed. The structured-programming concept was also embodied in new computer languages, such as Pascal, invented in 1971, which became the most popular language used in undergraduate programming education for twenty years. Structured programming was also one of the key concepts embodied in the Ada programming language commissioned by the U.S. Department of Defense for developing safety-critical software.

Another attack on the complexity problem was the use of formal methods. Here, an attempt was made to simplify and mathematize the design process by which programs were created. For technical and cultural reasons, formal methods failed to make a major impact. Technically, they worked well on relatively small programs, but failed to scale up to very large, complex programs. The cultural problem was that software practitioners—several years out of college, with rusty math skills—were intimidated by the mainly academic proponents of formal methods.

A much more widely adopted concept was the development model, which was as much a management tool as a technical one. The development model viewed the software writing process not as a once-and-for-all construction project, like the Hoover Dam, but as a more organic process, like the building of a city. Thus software would be conceived, specified, developed, and go into service—where it would be improved from time to time. After a period of use the software would die, as it became increasingly obsolete, and then the process would begin again. The different stages in the software life-cycle model made projects much easier to manage and control, and it recognized the ability of software to evolve in an organic way.

Software Products

Despite the improving software techniques that emerged toward the end of the 1960s, the gulf between the capabilities of computers to exploit software and its supply continued to widen. Except for the very largest corporations, it had become economically infeasible for most computer users to exploit the potential of their computers by writing very large programs because it would never be possible to recover the development costs. This was true whether they developed the programs themselves or had them written by a software contractor.

This situation created an opportunity for existing software contractors to develop software products—packaged programs, whose cost could be recovered through sales to ten or even a hundred customers. The use of a package was much more cost-effective than custom software. Sometimes a firm would tailor the package to its business, but as often it would adjust its business operations to take advantage of an existing product.

After the launch of the IBM System/360, which established the first stable industry-standard platform, a handful of software contractors began to explore the idea of converting some of their existing software artifacts into packages. In 1967 the market for software packages was still in its infancy, however, with fewer than fifty products available.

The development of the packaged-software industry was accelerated by IBM's unbundling decision of December 1968, by which it decided to price its hardware and software separately. Prior to this time IBM had, in line with its total-systems approach, supplied hardware, software, and system support, all bundled as a complete package. The user paid for the computer hardware and got the programs and customer support for "free." This was the standard industry practice at the time and was adopted by all of IBM's mainframe competitors.

It has never been established whether ordinary commercial judgment or antitrust pressure lay behind the unbundling decision. According to IBM sources, the decision was made because of rising software development costs. Between 1960 and 1969, the software component of IBM's R&D costs rose from about one-twentieth to about one-third of the total. IBM's customers paid for this software whether they used it or not, and the decision is said to have been made to release these customers from this "onerous" obligation.

It took about three years for the software products market to become fully established after unbundling. No doubt a thriving packaged-software industry would have developed in the long run, but the unbundling decision accelerated the process by transforming almost overnight the common perception of software from being a free good to becoming a tradable commodity. For example, while in the 1960s the American insurance industry had largely made do with IBM's free-of-charge software, unbundling was "the major event triggering an explosion of software firms and software packages for the life insurance industry." By 1972 there were eighty-one vendors offering 275 packages just for the life insurance industry. Several computer services firms and major computer users also recognized the opportunity to recover their development costs by spinning off software products. Boeing, Lockheed, and McDonnell Douglas had all developed engineering-design and other software at great expense.

Perhaps the most spectacular beneficiary of the new environment for software products was Informatics, the developer of the top-selling Mark IV file management system, one of the first database products. Informatics was founded in 1962 as a regular software contractor. In 1964, however, the company recognized that the database offerings of the mainframe manufacturers were very weak and saw a niche for a for-sale product. It took three years and cost about $500,000 to develop Mark IV. When the product was launched in 1967, there were few precedents for software pricing; its purchase price of $30,000 "astounded" computer users, long used to obtaining software without any separate charge. By late 1968 it had only a modest success, with 44 sales. After unbundling, however, growth was explosive—170 installations by the spring of 1969, 300 by 1970, and 600 by 1973. Mark IV took on a life of its own, and even had its own user group—called the IV (pronounced "ivy") League. Mark IV was the world's most successful software product for a period of fifteen years, until 1983, by which time it had cumulative sales of over $100 million.

By the mid-1970s all the existing computer manufacturers were important members of the software products industry. IBM was, of course, a major player from the day it unbundled. Despite its often mediocre products, it had two key advantages: inertia sales to the huge base of preexisting users of its software and the ability to lease its software for a few hundred dollars per month. The latter factor gave it a great advantage over the rest of the software industry, which needed outright sales to generate cash flow.

The most successful software products supplier of the late 1970s and 1980s was Computer Associates. Formed in 1976, the company initially occupied a niche supplying a sorting program for IBM mainframes. Computer Associates was one of the first computer software makers to follow a strategy of growth by buying out other companies. All Computer Associates' acquisitions, however, were aimed at acquiring software assets—"legitimate products with strong sales"—rather than the firms themselves; indeed, it usually fired half the staff of the companies it acquired. Over the next fifteen years, Computer Associates took over more than two dozen software companies, including some of the very largest. By 1989 it had annual revenues of $1.3 billion and was the largest independent software company in the world. This position did not last for long, however. By the 1990s, the old-line software products industry was being eclipsed by Microsoft and the other leading personal-computer software companies.

9

:: NEW MODES OF COMPUTING

BY THE mid-1960s, data-processing computers for business had become well established. The commercial computer installation was characterized by a large, centralized computer manufactured by IBM or one of the other half-dozen mainframe computer companies, running a batch-processing or real-time application. The role of the user in this computing environment was to feed data into the computer system and interact with it in the very restricted fashion determined by the application—whether it was an airline reservations system or an Automatic Teller Machine. In the case of the latter, users were probably unaware that they were using a computer at all. In a period of twenty years, computer hardware costs had fallen dramatically, producing ever greater computing power for the dollar, but the nature of data processing had changed hardly at all. Improved machines and software enabled more sophisticated applications and many primitive batch-processing systems became real time, but data processing still consisted of computing delivered to naive users by an elite group of systems analysts and software developers.

Today, when most people think of computing, they think of the personal computer on their desks, and it is hard to see a relationship between this kind of computing and commercial data processing. The reason is that there *is* no relationship. The personal computer grew out of an entirely different culture of computing, which is the subject of this chapter. This other mode of computing is associated with computer time-sharing, the BASIC programming language, Unix, minicomputers, and new microelectronic devices.

The Compatible Time-Sharing System

A time-sharing computer was one organized so that it could be used simultaneously by many users, each person having the illusion of being the sole user of the system—which, in effect, became his or her personal machine.

The first time-sharing computer system originated at MIT in 1961, initially as a way of easing the difficulties that faculty and students were beginning to encounter when trying to develop programs. In the very early days of computing, when people were using the first research machines at MIT and elsewhere, it was usual for a user to book the computer for a period of perhaps half an hour or an hour, during which he (occasionally she) would run a program, make corrections, and, it was hoped, obtain some results. In effect, the computer was the user's personal machine during that interval of time.

This was a very wasteful way to use a computer, however, because users would often spend a lot of time simply scratching their heads over the results and might get only a few minutes of productive computing done during the hour they monopolized the machine. Besides being an uneconomical way to use a machine costing perhaps $100 an hour to run, it was also an immense source of frustration to the dozens of people competing to use the machine, for whom there were simply not enough hours of hands-on access in the day. Up until the early 1960s, this problem had been addressed by batch-processing operating regimens. In a batch-processing system, in order to get the maximum amount of work from an expensive computer, it was installed in a "computer center"; users were required to punch their programs onto cards and submit them to the computer, not directly, but through the center's reception desk. In the computer center, a team of operators would put several programs together (a batch) and run them through the machine in quick succession. Users would then collect their results later the same day or the following day.

When the Department of Electrical Engineering at MIT got its first commercially manufactured computer, an IBM 704, in 1957, it introduced a batch-processing regimen. While batch processing made excellent use of the machine's time, faculty and researchers found it made very poor use of their time. They had to cool their heels waiting for programs to percolate through the system, which meant they could test and run their programs only once or twice a day. A complex program might take several weeks to debug before any useful results were obtained. Much the same was happening at other universities and research organizations around the world.

A possible solution to this problem, time-sharing, was first described by a British computer scientist named Christopher Strachey in 1959 when he proposed (and patented) the idea of attaching several operating consoles, each complete with a card reader and a printer, to a single mainframe computer. With some clever hardware and software, it would then be possible for all the users simultaneously to use the expensive mainframe computer, which would be sufficiently fast that users would be unaware that they were sharing it with others. At MIT John McCarthy, also a pioneer in artificial intelligence, came up with a similar, though more intimate, idea: users of the time-sharing system would communicate through typewriter-like terminals rather than the consoles that Strachey had envisaged.

The computing center at MIT was the first to implement a time-sharing system, in a project led by Robert Fano and Fernando Corbato. A demonstration version of MIT's CTSS—the Compatible Time-Sharing System—was shown in November 1961. This early, experimental system allowed just three users to share the computer, enabling them independently to edit and correct programs, and to do other information-processing tasks. For these users it was as though, once again, they had their own personal machine. Time-sharing set the research agenda for computing at MIT and many other universities for the next decade. Within a year of MIT's successful demonstration, several other universities, research organizations, and manufacturers had begun to develop their own time-sharing computers.

BASIC

The best known of these systems was the Dartmouth Time-Sharing System (DTSS). While MIT's CTSS was always a system for computer scientists—whatever the intentions of its designers—the one at Dartmouth College was intended for use by a much broader spectrum of users. The design of the system began in 1962, when a mathematics professor, John Kemeny (later the college's president), and Thomas E. Kurtz of the college computer center secured funding to develop a simple time-sharing computer system. They evaluated several manufacturers' equipment and eventually chose General Electric computers because of the firm's stated intentions of developing time-sharing systems. Kemeny and Kurtz took delivery of the computers at the beginning of 1964. General Electric had made some hardware modifications to support time-sharing, but there was no software provided at all—Kemeny and Kurtz would have to develop their own. They therefore decided to keep the system very simple so that it could be entirely developed by a group of student programmers. The group developed both the operating system and a simple programming language called BASIC in the spring of 1964.

BASIC—the Beginners All-purpose Symbolic Instruction Code—was a very simple programming language, designed so that undergraduates, liberal arts as well as science students, could develop their own programs. Until BASIC was developed, most undergraduates were forced to write their programs in FORTRAN, which had been designed for an entirely different context than education—for scientists and engineers to write technical computer applications. FORTRAN contained many ugly and difficult features. For example, to print out a set of three numbers, one was obliged to write something like:

$$\text{WRITE (6, 52) A, B, C}$$
$$\text{52 FORMAT (1H , 3F10.4)}$$

Another problem with FORTRAN was that it was very slow—even quite short programs would take several minutes to translate on a large IBM mainframe. For industrial users this was acceptable, as they tended to write production programs that were translated once and then used many times. But Kemeny and Kurtz rejected FORTRAN as a language for Dartmouth College's liberal arts students. They needed a new language that was

> simple enough to allow the complete novice to program and run problems after only several hours of lessons. It had to respond as expected to most common formulas and constructions. It had to provide simple yet complete error messages to allow the nonexpert to correct his program quickly without consulting a manual. It also had to permit extensions to carry on the most sophisticated tasks required by experts.

The language that Kemeny designed had most of the power of FORTRAN but expressed in a much easier-to-use notation. For example, to print out three numbers a programmer would simply write:

$$\text{PRINT A, B, C}$$

Starting in April 1964, Kemeny, Kurtz, and a dozen undergraduate students began to write a BASIC compiler and the operating software for the new time-sharing system. The fact that a small team of undergraduate programmers could implement a BASIC translator was itself a remarkable testament to the unfussy design and the simplicity of the language. By contrast, the big programming languages then being developed by the mainframe computer manufacturers were taking 50 or 100 programmer-years to develop.

Dartmouth BASIC became available in the spring of 1964, and the following academic year incoming freshmen began to learn to use the language in their foundation mathematics course. BASIC was so simple that it required just two one-hour lectures to get novices started on programming. Eventually most students, of whom only about a quarter were science majors, learned to use the system—typically for homework-type calculations, as well as for accessing a large range of library programs, including educational packages, simulations, and computer games. By 1968 the system was being used by twenty-three local schools and ten New England colleges.

BASIC was carried along on the crest of the wave of time-sharing systems that were coming into operation in the late 1960s and early 1970s. Every computer manufacturer had to provide a BASIC translator if it was to compete in the educational market. Very soon, BASIC became the introductory language for almost all nonspecialist students learning computing. In 1975 it became the first widely available programming language for the newly emerging personal computer and laid the foundations of Microsoft.

Computer scientists have often criticized BASIC, seeing it as a step backward in terms of software technology. This may have been true from a technical viewpoint, but most critics overlooked BASIC's much more important cultural significance in establishing a user-friendly programming system that enabled ordinary people to use computers without a professional computer programmer as an intermediary. Before BASIC came on the scene there were two major constituencies of computer users: computer professionals who developed applications for other people and naive computer users, such as airline reservations clerks, who operated computer terminals in a way that was entirely prescribed by the software. BASIC created a third group: users who could develop their own programs and for whom the computer was a personal information tool.

J. C. R. Licklider and the Advanced Research Projects Agency

Up to the mid-1960s, although the time-sharing idea caught on at a few academic institutions and research organizations, the great majority of computer users continued to use computers in the traditional way. What propelled time-sharing into the mainstream of computing was the lavish funding to develop systems that the Advanced Research Projects Agency (ARPA) began to provide in 1962. ARPA was one of the great, but little-known, cultural forces in shaping U.S. computing.

ARPA was initially established as a response to the Sputnik crisis of October 1957, when the Soviet Union launched the first satellite. This event caused something close to panic in political and scientific circles by throwing into question the United States' dominance of science and technology. President Eisenhower responded with massive new support for science education and scientific research, which included ARPA to sponsor and coordinate defense-related research. ARPA was not required to deliver immediate results, but to pursue goals that would have a long-term payoff.

Among ARPA's projects in 1962 was a $7 million program to promote the use of computers in defense. The person appointed to direct this program was an MIT psychologist and computer scientist, J. C. R. Licklider. Insofar as today's interactive style of computing can be said to have a single parent, that parent was Licklider. The reason for his achievement was that he was a psychologist first and a computer scientist second.

J. C. R. "Lick" Licklider studied psychology as an undergraduate, during the war worked at Harvard University's Acoustical Laboratory, and stayed on as a lecturer. In 1950 he accepted a post at MIT and established the university's psychology program in the Department of Electrical Engineering, where he was convinced it would do the most good, by encouraging engineers to design with human beings in mind. For a period in the mid-1950s he was involved with the SAGE defense project, where he did important work on the human-factors design of radar display consoles. In 1957 he became a vice president of the Cambridge-based R&D firm Bolt, Beranek, and Newman (BBN), where he formulated a manifesto for human-computer interaction, published in 1960. His classic paper, *Man-Computer Symbiosis,* was to reshape computing over the next twenty years.

The single most important idea in Licklider's paper was to advocate the use of computers to augment the human intellect. This was a radical viewpoint at the time and, to a degree, at odds with the prevailing view of the computer establishment, particularly artificial intelligence (AI) researchers. Many computer scientists were caught up by the dream that AI would soon rival the human intellect in areas such as problem solving, pattern recognition, and chess playing. Soon, they believed, a researcher would be able to assign high-level tasks to a computer. Licklider believed that such views were utopian.

He argued that computer scientists should develop systems that enabled people to use computers to enhance their everyday work. He believed that a researcher typically spent only 15 percent of his or her time "thinking"—the rest of the time was spent looking up facts in reference books, plotting graphs, doing calculations, and so on. It was these low-level tasks that he wanted computers to automate instead of pursuing a high-flown vision of AI. He argued that it might be twenty years before computers could do useful problem-solving tasks: "That would leave, say, five years to develop man-computer symbiosis and fifteen years to use it. The fifteen may be ten or 500, but those years should be intellectually the most creative and exciting in the history of mankind."

Licklider was elaborating these ideas at BBN in a project he called the *Library of the Future* when he was invited in 1962 to direct a program at ARPA. Taking a wide view of his mandate, he found at his disposal the resources to pursue his vision of man-computer symbiosis.

A consummate political operator, Licklider convinced his superiors to establish within ARPA the Information Processing Techniques Office (IPTO) with a budget that eventually exceeded that of all other sources of U.S. public research funding for computing combined. Licklider developed an effective way of administrating the IPTO program, which was to place his trust in a small number of academic centers of excellence that shared his vision—including MIT, Stanford, Carnegie-Mellon, and Utah—and give them the freedom to pursue long-term research goals with a minimum of interference. The vision of interactive computing required research advances along many fronts: computer graphics, software engineering, and the psychology of human-computer interaction. By producing a stream of new technologies of wide applicability, Licklider kept his paymasters in the government happy.

Early on, Licklider encouraged his chosen centers to develop time-sharing systems that would facilitate studies of human-computer interaction. The centerpiece of the time-sharing programs was a $3 million grant made to MIT to establish a leading-edge system, Project MAC—an acronym that was "variously translated as standing for multiple-access computer, machine-aided cognition, or man and computer." Based on IBM mainframes, the system became operational in 1963 and eventually built up to a total of 160 typewriter consoles on campus and in the homes of faculty. Up to thirty users could be active at any one time. Project MAC allowed users to do simple calculations, write programs, and run programs written by others; it could also be used to prepare and edit documents, paving the way for word-processing systems. During the summer of 1963, MIT organized a summer school to promote the time-sharing idea, particularly within the mainframe computer companies. Fernando Corbato recalled:

> The intent of the summer session was to make a splash. We were absolutely frustrated by the fact that we could not get any of the vendors to see a market

in this kind of machine. They were viewing it just as some special purpose gadget to amuse some academics.

The summer school had a mixed success in changing attitudes. While the degree of interest in time-sharing was raised in the academic community, IBM and most of the other conservative mainframe computer manufacturers remained unmoved. It would be a couple of years before the climate for time-sharing would start to change.

By 1965 Project MAC was fully operational but was so heavily overloaded that MIT decided, with ARPA funding, to embark on an even bigger time-sharing computer system that would be able to grow in a modular way and eventually support several hundred users. This system would be known as Multics (for Multiplexed Information and Computing Service). MIT had a long-standing relationship with IBM, and it had been assumed by the senior administrators in both organizations that MIT would choose to use an IBM machine again. At the grassroots level in MIT, however, there was growing resistance to choosing another IBM machine. IBM seemed complacent, and its mainframes were technologically unsuited to time-sharing; MIT decided to choose a more forward-looking manufacturer.

Unfortunately for IBM, when it had designed its new System/360 computers in 1962, time-sharing was barely on the horizon. Already overwhelmed by the teething problems of its new range of computers, the company was unwilling to modify the design to fulfill MIT's needs. By contrast, General Electric, which had already decided to make a time-sharing computer based on the Dartmouth system, was happy not only to collaborate with MIT, but to have it directly participate in the specification of its new model 645 computer, which was to be designed from the ground up as a dedicated time-sharing computer. It was an opportunity for GE to make a last effort to establish a profitable mainframe operation by attacking a market where IBM was particularly weak.

Multics was to be the most ambitious time-sharing system yet; costing up to $7 million, it would support up to 1,000 terminals, with 300 in use at any one time. To develop the operating software for the new system, MIT and General Electric were joined by Bell Labs. Bell Labs became the software contractor because the company was rich in computing talent but was not permitted, as a government-regulated monopoly, to operate as an independent computer-services company. The Multics contract enabled Bell Labs to find an outlet for its computing expertise and to continue building up its software writing capabilities.

Having been pushed aside by Project MAC, IBM's top management realized that the company could bury its head in the sand no longer and would have to get onto the time-sharing scene. In August 1966 it announced a new member of the System/360 family, the time-sharing model 67. By this time most other computer manufacturers had also tentatively begun to offer their first time-sharing computer systems.

Computer Utility

The development of the first time-sharing computer systems attracted a great deal of attention both inside and outside the computer-using community. The original idea of time-sharing was broadened into the "computer utility" concept.

The idea of a computer utility originated at MIT around 1964 as an analogy to the power utilities. Just as the ordinary electricity user drew power from a utility, rather than owning an independent generating plant, so, it was argued, computer users should be served by giant centralized mainframes from which they could draw computing power, instead of owning individual computers. Although some conservative forces regarded the computer utility as a fad, most computer users were caught up by the vision. Time-sharing and the idea of the computer utility became the "hottest new talk of the trade" in 1965. Skeptics remained, but they were a minority. One correspondent of the *Datamation* computer magazine, for example, remarked that he was "amazed" by the way the computer-using community had "overwhelmingly and almost irreversibly committed itself emotionally and professionally to time-sharing" and was being seduced by the prestige of MIT and Project MAC into "irrational, bandwagon behavior."

Even if the skeptics were right (and so they were proved), the appeal of the computer utility was seemingly unstoppable. The computer and business press regularly featured articles on the subject, and books appeared with titles such as *The Challenge of the Computer Utility* and *The Future of the Computer Utility*. *Fortune* magazine reported that computer industry prophets were convinced that the computer business would "evolve into a problem-solving service . . . a kind of computer utility that will serve anybody and everybody, big and small, with thousands of remote terminals connected to the appropriate central processors."

There were two key forces driving the computer utility idea, one economic and one visionary. Computer manufacturers and buyers of computer equipment were primarily interested in the economic motive, which was underpinned by "Grosch's Law." Herb Grosch was a prominent computer pundit and industry gadfly of the period whose claims to fame included a flair for self-publicity and an ability to irritate IBM; history probably would not remember him except for his one major observation. Grosch's Law quantified the economies of scale in computing by observing that the power of a computer varied as the square of its price. Thus a $200,000 computer would be approximately four times as powerful as one costing $100,000; while a $1 million computer would be twenty-five times as powerful as the $200,000 model. Because of Grosch's Law, it was clear that it would be much cheaper to have twenty-five users sharing a single $1 million computer than to provide each of them with their own $200,000 computer. Without this economic motive, there would have been no financial incentive for time-sharing and no time-sharing industry.

The visionaries of the computer utility, however, were caught up by a broader aim for the democratization of computing. Martin Greenberger, a professor in the MIT Management School who was probably the first to describe the utility concept in print, argued in an *Atlantic Monthly* article that the drive for computer utilities was unstoppable and that "[b]arring unforeseen obstacles, an on-line interactive computer service, provided commercially by an information utility, may be as commonplace by A.D. 2000 as the telephone service is today."

The computer utility vision was widely shared by computer pundits at the end of the 1960s. Paul Baran, a computer-communications specialist in the RAND Corporation, waxed eloquent about the computer utility in the home:

> And while it may seem odd to think of piping computing power into homes, it may not be as far-fetched as it sounds. We will speak to the computers of the future in a language simple to learn and easy to use. Our home computer console will be used to send and receive messages—like telegrams. We could check to see whether the local department store has the advertised sports shirt in stock in the desired color and size. We could ask when delivery would be guaranteed, if we ordered. The information would be up-to-the-minute and accurate. We could pay our bills and compute our taxes via the console. We would ask questions and receive answers from "information banks"— automated versions of today's libraries. We would obtain up-to-the-minute listing of all television and radio programs. We could use the computer to preserve and modify our Christmas lists. It could type out the names and addresses for our envelopes. We could store in birthdays. The computer could, itself, send a message to remind us of an impending anniversary and save us from the disastrous consequences of forgetfulness.

Using computers to remind the user about impending birthdays and anniversaries was a vacuous suggestion common to many home-computing scenarios of the period. This suggests a paucity of ideas for the domestic uses of computers that was never really addressed in the stampede toward time-sharing.

By 1967 time-sharing computers were popping up all over the country—with twenty firms competing for a market estimated at $15 million to $20 million a year. IBM was offering a service known as QUICKTRAN for simple calculations in five cities, with plans to double that number. General Electric had established time-sharing computers in twenty U.S. cities, and Licklider's old employer, BBN, had established a nationwide service called Telcomp. Other new entrants were Tymshare in San Francisco, Keydata in Boston, and Comshare in Ann Arbor, Michigan. One of the biggest time-sharing operations was that of the Dallas-based University Computing Company (UCC), which established major computer utilities in New York and Washington. By 1968 it had several computer centers with terminals located in thirty states and a dozen countries. UCC was one of the glamour stocks of the 1960s; during 1967–68 its stock price rose from $1.50 to $155. By the end of the 1960s, the major time-sharing firms had expanded across North America and into Europe.

Then, almost overnight, the bottom dropped out of the computer utility market. As early as 1970 the computer utility was being described by some industry insiders as an "illusion" and one of the "computer myths of the 60's." By 1971 several firms were in trouble. UCC saw its share price dive from a peak of $186 to $17 in a few months. The demise of the time-sharing industry was to set back the computer utility dream for the best part of twenty years. While simple time-sharing computer systems survived, the broad vision of the computer utility was killed off by two unforeseen and unrelated circumstances: the software crisis and the fall in hardware prices.

Big computer utilities were critically dependent on software to make them work. The more powerful the computer utility, the harder it became to write the software. IBM was the first to discover this. The company was already embroiled in the disastrous OS/360 software development for its new System/360 mainframes, and launching the time-sharing model 67 involved it in another major challenge to develop time-sharing software. Originally due for delivery in the fall of 1966, the schedule slipped again and again until IBM ended up with a "team of hundreds . . . working two and a half shifts trying to get the software ready." By the time the first late and unreliable release of the software appeared in 1967, many of the original orders had been canceled. It was estimated that IBM lost $50 million in this foray into time-sharing.

General Electric was even more badly hit by the software crisis. Having succeeded so easily with the relatively simple Dartmouth-based system, and with thirty time-sharing centers in the United States and Europe running commercially,

GE expected development of the new Multics system to be straightforward. It and its partners, MIT and Bell Labs, were wholly unprepared for the magnitude of the software development problems. After four years of costly and fruitless development, Bell Labs finally pulled out of the project in early 1969, and it was not until the end of that year that an early and rather limited version of the system limped into use at MIT. In 1970 General Electric decided to withdraw its model 645 computer from the market, fearing that the software problems were simply insuperable; a few months later it pulled out of the mainframe computer business altogether.

The firms that remained in the time-sharing business realized that, because of software problems, for the foreseeable future only small, specialized systems would be viable. Those systems, typically with thirty to fifty users, mainly supplied niche markets for science, engineering, and business calculations and were a pale shadow of the vision of the computer utility. Nonetheless, despite the failure of the computer utility, there remained a latent desire for communal computing that would surface again with the Internet in the 1990s.

Unix

There was, however, one very beneficial spin-off of the Multics imbroglio: the Unix operating system, developed at Bell Labs during 1969–74. The software disasters epitomized by OS/360 and Multics occurred because they were too baroque. This led a number of systems designers to explore a small-is-beautiful approach to software that tried to avoid complexity by a clear minimalist design. The Unix operating system was the outstanding example of this approach. It was the software equivalent of the Bauhaus chair.

Unix was developed by two of Bell Labs' programmers, Ken Thompson and Dennis M. Ritchie, both of whom—bearded, long-haired, and portly—conformed to the stereotype of the 1960s computer guru; their appearance has been little changed by time. When Bell Labs pulled out of the Multics project in 1969, Thompson and Ritchie were left frustrated because working on Multics had provided them with a rather attractive programming environment:

> We didn't want to lose the pleasant niche we occupied, because no similar ones were available; even the time-sharing service that would later be offered under GE's operating system did not exist. What we wanted to preserve was not just a good environment in which to do programming, but a system around which a fellowship could form.

Thompson and Ritchie had a free rein at Bell Labs to pursue their own ideas and decided to explore the development of a small, elegant operating system that they later called Unix—"a somewhat treacherous pun on 'Multics.'" Hardware was less forthcoming than freedom to pursue this research whim, however, and they "scrounged around, coming up with a discarded obsolete computer"—a PDP-7 manufactured by the Digital Equipment Corporation (PDP stands for "programmed data processor"). The PDP–7 was a small computer designed for dedicated laboratory applications and provided only a fraction of the power of a conventional mainframe. The design of Unix evolved over a period of a few months in 1969 based on a small set of primitive concepts. (One of the more elegant concepts was the program "pipe," which enabled the output from one program to be fed into another, rather like using pipework in an industrial production process. Using pipes, very complex programs could be constructed out of simpler ones.)

By early 1970 Unix was working to its creators' satisfaction, providing remarkably powerful facilities for a single user on the PDP–7. Even so, Thompson and Ritchie had a hard time convincing their colleagues of its merits. Eventually, by offering to develop some text-processing software, they coaxed the Bell Labs patent department into using the system for preparing patent specifications. With a real prospective user, funding was obtained for a larger computer, and the newly launched Digital Equipment PDP–11/45 was selected. The system was completely rewritten in a language called "C," which was designed especially for the purpose by Ritchie. This was a "systems implementation language," designed for writing programming systems in much the same way that FORTRAN and COBOL were designed for writing scientific and commercial applications. The use of C made Unix "portable," so that it could be implemented on any computer system—a unique operating-system achievement at the time. (Soon C, like Unix, was to take on a life of its own.)

Unix was well placed to take advantage of a mood swing in computer usage in the early 1970s caused by a growing exasperation with large, centralized mainframe computers and the failure of the large computer utility. A joke circulating at the time ran:

Q: What is an elephant?
A: A mouse with an IBM operating system.

Independent-minded users were beginning to reject the centralized mainframe in favor of a decentralized small computer in their own department. Few manufacturers offered suitable operating systems, however, and Unix filled the void. Ritchie recalled: "Because they were starting afresh, and because manufacturers' software was, at best, unimaginative and often horrible, some adventuresome people were willing to take a chance on a new and intriguing, even though unsupported, operating system." This probably would have happened irrespective of Unix's merits, since there was little else available and Bell Labs had adopted the benign policy of licensing the system to colleges and universities at a nominal cost. But the minimalist design of Unix established an immediate rapport with the academic world and research laboratories. From about 1974, more and more colleges and universities began to use the system; and within a couple of years graduates began to import the Unix culture into industry.

A major change in the Unix world occurred in 1977, when it began to grow organically; more and more software was added to the basic system developed by Thompson and Ritchie. The clean, functional design of Unix made this organic growth possible without affecting the inherent reliability of the system. Very powerful versions of Unix were developed by the University of California at Berkeley, the computer workstation manufacturer Sun Microsystems, and other computer manufacturers. In much the way that FORTRAN had become a standard language by default in the 1950s, Unix was on the way to becoming the standard operating system for the 1980s. In 1983 Thompson and Ritchie received the prestigious ACM Turing Award, on which it was stated: "The genius of the UNIX system is its framework, which enables programmers to stand on the work of others." Unix is one of the design masterpieces of the twentieth century.

In the late 1990s—between the appearance of the first and second editions of this book—Unix took on yet another mantle as the Linux operating system for personal computers. Originally started by a Finnish computer science undergraduate Linus Torvalds in 1992, the operating system was perfected and extended by contributions of countless individual programmers, working without direct payment, often on their own time. Linux has become one of the centerpieces of the "open-source movement" (discussed further in chapter 12).

Minicomputers

Returning to the demise of the computer utility, if the software crisis had not been enough to kill the concept unaided, then falling hardware prices during the 1970s would have done so. Between 1965 and 1975, the introduction of integrated circuit electronics reduced the cost of computer power by a factor of a hundred, undermining the prime economic justification for time-sharing—that is, sharing the cost of a large computer by spreading it across many users. By

1970 it was possible to buy for around $20,000 a "minicomputer" with the power of a 1965 mainframe that had cost ten times as much. Rather than subscribe to a time-sharing service, which cost upward of $10 per hour for each user, it made more economic sense for a computer user to buy outright a small time-sharing system that supported perhaps a dozen users. During the 1970s the small in-house time-sharing system became the dominant mode of computing in universities, research organizations, and many businesses.

The minicomputer did not come from the established mainframe computer industry, however, but from Project Whirlwind and MIT—via the electronics industry. The minicomputer was part of a much bigger "revolution in miniature" that swept the U.S. electronics industry in the 1950s and 1960s. After producing the technology for minicomputers, the revolution went on to produce hand-held electronic calculators, digital watches, video games, and eventually the personal computer.

The period from 1950 to 1965 saw three transformations in the electronics industry as the technology evolved from vacuum tubes to discrete transistors and finally to integrated circuits (or "chips"). These developments were largely independent of the established computer industry, although computers appropriated the new electronics technologies as they became available—and this led to the popular classification of computers as first-, second-, and third-generation as they incorporated in turn each of the three major electronics innovations. For the most part, however, the two industries pursued independent paths.

The U.S. electronics industry of the 1950s and 1960s was highly concentrated in two main areas: around Route 128 near Boston and in Silicon Valley in the San Francisco Bay Area. The first firms were located in these regions because of their proximity to major research universities (MIT and Stanford). In addition, both areas had well-developed venture-capital firms that were willing to fund risky, high-tech enterprises and provide mature counsel to their young engineer founders. As the number of firms grew from a handful to a few dozen in the 1960s, and eventually to hundreds in the 1970s, new firms tended to cluster in the same areas. The new firms, besides often being formed by people already working in the area, benefited from nearby suppliers to the microelectronics industry, the availability of a highly trained workforce, and the rapid diffusion of know-how through a network of social relationships. This was the context in which the Digital Equipment Corporation, the leading manufacturer of minicomputers, was established.

The Digital Equipment Corporation (DEC) was formed by Kenneth Olsen and Harlan Anderson in 1957. Olsen, then aged thirty-one, had graduated in electrical engineering from MIT in 1950, after which he became a research associate on Project Whirlwind—where he did extensive work turning the prototype core-memory development into a reliable system. Although Olsen retained strong links with MIT, he was more excited by turning the technologies into real products than in academic research. In 1957 he secured $70,000 of venture capital from American Research and Development (ARD).

ARD was founded in 1946 by a Harvard Business School professor, General George F. Doriot—the "father of venture capital"—to finance the commercial exploitation of technologies developed during the war. ARD was the prototype venture-capital firm, and the development of such financial operations was a key factor responsible for the dynamism of the new high-tech industries in the United States. Most overseas countries found it very difficult to compete with U.S. firms until they established their own venture-funding organizations.

Olsen's aim was to go into the computer business and compete with the mainframe manufacturers. However, in the late 1950s this was not a realistic short-term goal. The barriers to entry into the mainframe business were rising. In order to enter the mainframe business one needed three things, in addition to a central processing unit: peripherals (such as magnetic tape and disk drives), software (both applications and program development tools), and a sales force. It would cost several hundred million dollars to establish all these capabilities. Because of these formidable barriers to entering the computer industry, Doriot convinced Olsen to first establish the firm with more attainable objectives.

DEC set up operations in part of a pre–Civil War woolen mill in Maynard, Massachusetts, close to the Route 128 electronics industry. For the first three years of its existence, the company produced digital circuit boards for the booming digital electronics industry. This proved to be a successful niche and provided the funding for DEC's first computer development. DEC announced its first computer, the PDP–1, in 1960. Olsen chose the term "programmed data processor" because the new computer cost $125,000 for a basic model and existing computer users "could not believe that in 1960 computers that could do the job could be built for less than $1 million."

How was it possible for DEC to produce a computer for a fifth or a tenth of what it cost the mainframe computer industry? The answer is that in a mainframe computer, probably only 20 percent of the cost was accounted for by the central processor—the rest went on peripherals, software, and marketing. Olsen aimed the PDP–1 not at the commercial data-processing user, however, but at the science and engineering market. These customers did not need advanced peripherals, they were capable of writing their own software, and they did not need a professional sales engineer to analyze their applications. The PDP–1 and half a dozen other models produced over the next five years were a modest success, though not sufficient to reshape the industry. This would happen with the introduction of the PDP–8 minicomputer in 1965. The PDP–8 was one of the first computers to exploit the newly emerging technology of integrated circuits, and as a result it was far smaller and cheaper than any of DEC's previous computers. The PDP–8 would fit in a packing case and sold for just $18,000.

The PDP–8 was an instant success, and several hundred systems were delivered during the following year. The success of the PDP–8 enabled DEC to make its first public offering in 1966. By that time the firm had sold a total of 800 computers (half of them PDP–8s), employed 1,100 staff, and occupied the entirety of its Maynard woolen-mill headquarters. The offering netted $4.8 million—a handsome return on ARD's original investment. Over the next ten years, the PDP–8 remained constantly in production, eventually selling between 30,000 and 40,000 systems. Many of these new computers were used in dedicated applications, such as factory process automation, where a traditional computer would have been too expensive. Other systems were sold directly to other engineering corporations for inclusion in advanced instruments such as medical scanners.

Many PDP–8s found their way into colleges and research laboratories, where their low price enabled research students and faculty to experience hands-on computing in a way that had not been possible since the 1950s. The result was, in many cases, to redirect the careers of some of these people into computing, when they had never previously considered such a career. Many of the users of PDP–8s became very attached to them, regarding them as their "personal" computers. Some users developed games for the machines—one of the most popular was a simulation of a moon-landing vehicle that the user had to guide to a safe landing. The experience of hands-on computing produced a strong computer hobbyist culture, not only among students and young technicians but also in the community of seasoned engineers.

One such engineer, Stephen Gray, editor of *Electronics* magazine, founded the Amateur Computer Society (ACS) in May 1966 in order to exchange information with other hobbyists about building computers. Drawing support from a widely scattered community of technically trained enthusiasts working in universities, defense firms, and electronics and computer companies, Gray began circulating his *ACS Newsletter* to 160 members in 1966. It presented information on circuit design and component availability and compatibility, and it encouraged subscribers by printing stories of successful home-built computers. In general, however, owning a computer was not a reasonable aspiration for the average computer hobbyist, when the cheapest machine cost $10,000. But in the 1970s the latent desire of computer hobbyists to own computers was a powerful force in shaping the personal computer.

By 1969, when the term *minicomputer* first came into popular use, small computers were a major sector of the computer industry. DEC had been joined by several other minicomputer manufacturers, such as Data General (formed by a group of ex-DEC engineers) and Prime Computer—both of which became major international firms. Several of the established electronics and computer manufacturers had also developed minicomputer divisions, including Hewlett-Packard, Harris, and Honeywell. DEC, as the first mover in the minicomputer industry, always had by far the strongest presence, however. By 1970 it was the world's third-largest computer manufacturer, after IBM and the UNIVAC Division of Sperry Rand. When the aging General Doriot retired from ARD in 1972, its original stake was worth $350 million.

DEC found itself increasingly the target of lower-cost imitators, and so it erected defensive barriers by differentiating its products from those of its competitors by producing comprehensive computer systems. It began to develop peripherals and sophisticated software packages, and to employ a large sales force. In 1967 DEC introduced its most powerful computer to date—the PDP–10, which was effectively a mainframe computer costing between half a million and one and a half million dollars. DEC developed time-sharing software for the computer and quickly established it as one of the most popular campus computing systems. When DEC introduced its PDP–11 computers in the early 1970s, however, it became increasingly difficult to use its time-sharing software as a way of differentiating its products from those of its competitors. Unix was becoming the preferred time-sharing operating system, making the particular computer on which it was run irrelevant.

West Coast Microelectronics

People often suppose that DEC grew out of the computer industry. This is not the case: It was in effect a computer firm born of MIT and the East Coast microelectronics industry. At the same time, Stanford University and the microelectronics industry of Silicon Valley were beginning to spin off digital electronics and computer firms on the other side of the country.

The West Coast microelectronics industry was established in the mid-1950s when William Shockley established Shockley Semiconductor Laboratories in Palo Alto, his hometown. Shockley was a distinguished scientist who had been a co-inventor of the transistor at Bell Laboratories in 1946, for which he shared a Nobel Prize. Shockley's commercial venture was a failure, however, and would have been of little importance except for eight of the people he recruited from the older electronics companies in the East. They were frustrated by Shockley's focus on discovering new devices instead of manufacturing them. In 1957 the "Shockley Eight" left the laboratory to form their own company, Fairchild Semiconductor, for which venture capital was provided by the Fairchild Camera and Instrument Company.

Fairchild Semiconductor spawned the phenomenon we now know as Silicon Valley. Most top managers in the Silicon Valley semiconductor firms spent some period of their careers in the company; at a conference in Sunnyvale, California, in 1969, it was said that of 400 engineers present, fewer than two dozen had never worked for Fairchild. In turn, Fairchild spun off many other start-up companies in the semiconductor industry, and these spun off others. The most important of these was Intel, formed in 1968, which developed semiconductor memories and the microprocessor (the latter will be described in the next chapter).

During the 1960s the pace of semiconductor innovation was furious. The first integrated circuits were produced in 1962 for the military. They cost about $50 and contained an average of half a dozen active components per chip. After that, the number of components on a chip doubled each year. By 1970 it was possible to make LSI chips (for Large Scale Integration) with a thousand active components. These were widely used in computer memories, and during the early 1970s decimated the core-memory industry. The rapidly increasing density of semiconductor chips was first commented upon by Gordon Moore—a co-founder of Intel—in 1965, when he observed that the number of components on a chip had "increased at a rate of roughly a factor of two per year" and that "this rate can be expected to continue for at least ten years." This projection became known as "Moore's Law," and he calculated "that by 1975, the number of components per integrated circuit will be 65,000." In fact, while the rate of doubling settled down at about eighteen months, the phenomenon persisted into the twenty-first century, by which time integrated circuits contained tens of millions of components.

Beyond the computer industry, integrated circuits and digital electronics enabled the development of a range of new consumer products.

First, the new chips transformed the calculator industry. The first electronic calculating machines, developed in the mid-1960s, used both off-the-shelf and custom chips—as many as a hundred in a machine. They were very expensive, and the old electromechanical technology remained competitive until the late 1960s. The early 1970s, however, saw a phenomenal improvement in calculator electronics as the number of chips in a calculator decreased from several dozen to several, and finally to a single chip. At this point the market for calculators switched from a low-volume one for business users to a high-volume one for educational and domestic users. The first hand-held calculators were introduced around 1971 at a price of $100; by 1975 this had fallen to under $5. The drop in prices drove many of the U.S. calculator firms out of business, leaving production to Japanese and Far East firms—although they remained dependent on U.S.-designed chips.

Another major development was the digital watch, which followed a similar boom-and-bust cycle. The first digital watches appeared on the market in 1973; they were virtually all U.S.-produced, and retailed for about $250. It used to be said (correctly) that a digital watch had approximately the calculating power of the ENIAC. By 1975 production had increased more than tenfold, and the price had dropped to $150. By 1980 world production—no longer dominated by the United States—had reached 100 million, and a watch cost a few dollars.

Another important user of the new high-density chips was the video game industry, of which the American firm Atari quickly emerged as the leader. Founded in 1971 by a thirty-eight-year-old entrepreneur, Nolan Bushnell, Atari's first product was an electronic table-tennis game called Pong (so called to avoid trademark infringement on the name Ping-Pong). Atari initially supplied its Pong game for use in amusement arcades, much as the rest of the amusements industry supplied pinball machines and jukeboxes. Its success spawned both U.S. and Japanese competitors, and video games became an important branch of the amusements industry. However, as the cost of chips plummeted, Bushnell saw an opportunity to move from low-volume production for arcades to high-volume production for the domestic market, much as had happened with calculators. Atari developed a domestic version of Pong, which plugged into the back of an ordinary TV set. The home version of Pong hit the stores in time for Christmas 1974 and sold for $350. Atari followed up with a string of domestic games, and by 1976 the firm had annual sales of $40 million, even though the price of a typical game had dropped to $50. By this time video games had become a major industry with many firms. Enormous sums of money were now needed to develop the next generation of games, so Bushnell decided to sell out to Time-Warner, which had sufficient money to develop the business. In 1978, Atari became a full-scale developer of personal computers.

It should be emphasized that none of these products—calculators, digital watches, or video games—was a computer in the true sense. Moreover, in the early 1970s, these electronics-based industries, and the traditional computer industry, were quite distinct sectors, with very little crossover between them. By the late 1970s, however, they would all converge around a single product—the personal computer, built around the microprocessor, which was a true computer.

Part Four

GETTING PERSONAL

10

:: THE SHAPING OF THE PERSONAL COMPUTER

NO HISTORIAN has yet written a full account of the personal computer, mainly because historians generally avoid writing about recent events on which they lack a proper perspective. But the personal computer has been of such sweeping global importance that no book on the history of the computer could properly ignore it.

There has, of course, been no shortage of published accounts of the development of the personal computer. Scores of books and hundreds of articles, written mostly by journalists, have appeared in response to a demand from the general public for an understanding of the personal computer. Much of this reportage is bad history, though some of it is good reading. Perhaps its most serious distortion is to focus on a handful of individuals, portrayed as visionaries who clearly saw the future and made it happen: Apple Computer's Steve Jobs and Microsoft's Bill Gates figure prominently in this genre. By contrast, IBM and the established computer firms are usually portrayed as dinosaurs: slow-moving, dim-witted, deservedly extinct. When it comes to be written, the history of the personal computer will be much more complex than this. It will be seen to be the result of a rich interplay of cultural forces and commercial interests.

Radio Days

If the idea that the personal computer was shaped by an interplay of cultural forces and commercial interests appears nebulous, it is useful to compare the development of the personal computer with the development of radio in the opening decades of the twentieth century, whose history is well understood. There are some useful parallels in the social construction of these two technologies, and an understanding of one can deepen one's understanding of the other.

In the 1890s the phenomenon we now call radio was a scientific novelty in search of an application. Radio broadcasting as we now know it would not emerge for a generation. The first commercial application of the new technology was in telegraphy, by which Morse signals were transmitted from one point to another—a telegraph without wires. Wireless telegraphy was demonstrated in a very public and compelling way in December 1901 when Guglielmo Marconi transmitted the letter S repeatedly in Morse across the Atlantic from Poldu in Cornwall, England, to St. John's, Newfoundland, Canada. Banner newspaper headlines reported Marconi's achievement, and his firm began to attract the attention of telegraph companies and private investors.

Over the next few years, wireless telegraphy was steadily perfected and incorporated into the world's telegraph systems, and voice transmission and marine-based telegraphs were developed. The latter, particularly, captured many headlines. In 1910 a telegraph message from the SS *Montrose* resulted in the capture of the "acid-bath murderer" Dr. Hawley Crippen as he fled from England to Canada. Two years later the life-saving role of the telegraph in the *Titanic* disaster resulted in legislation mandating that all ships holding fifty or more people must carry a permanently manned wireless station. All this media attention served to reinforce the dominant mode of the technology for the point-to-point transmission of messages.

While wireless telegraphy was in the process of being institutionalized by the telegraph companies and the government, it also began to draw the attention of men and boy hobbyists. They were attracted by the glamour associated with wireless telegraphy and by the excitement of "listening in." But mostly the amateurs were attracted by the technology itself—the sheer joy of constructing wireless "sets" and communicating their enthusiasm to like-minded individuals. By 1917 there were 13,581 licensed amateur operators in the United States, and the number of unlicensed receiving stations was estimated at 150,000.

The idea of radio broadcasting arose spontaneously in several places after World War I, although David Sarnoff (later of RCA) is often credited with the most definite proposal for a "radio music box" while he was working for the American Marconi Company in New York. Broadcasting needed an audience, and radio amateurs constituted that first audience. But for the existence of amateur operators and listeners, radio broadcasting might never have developed.

Once the first few radio stations were established, broadcasters and listeners were caught in a virtuous circle: More listeners justified better programs; and better programs enticed more listeners. Between 1921 and 1922, 564 radio stations came into existence in the United States. The flood fueled a demand for domestic radio receivers, and the radio-set industry was born. The leading firm, RCA, led by David Sarnoff, sold $80 million worth of radio sets in the four years beginning with 1921. Existing firms such as Westinghouse and General Electric also began to make radio sets, competing fiercely with start-up firms such as Amrad, De Forest, Stromberg-Carlson, Zenith, and many more. By the mid-1920s, the structure of American radio broadcasting had been fully determined, and it has been remarkably resilient to assaults from, in turn, cinema, television, satellite broadcasting, and cable.

Three key points emerge from this thumbnail history of American radio broadcasting. First, radio came from a new enabling technology whose long-term importance was initially unrecognized. Originally promoted as a point-to-point communications technology, radio was reconstructed into something quite different: a broadcast entertainment medium for the mass consumer. Second, a crucial set of actors in this transformation were the radio amateurs. They built the first receivers when there was no radio-set industry, thus enabling broadcasting to take off. They are the unsung heroes of the radio story. Finally, once radio broadcasting was established, it was quickly dominated by a few giant firms—radio-set manufacturers and broadcasters. Some of these firms were the creations of individual entrepreneurs, while others came from the established electrical engineering industry. Within a decade, the firms were virtually indistinguishable.

As we shall see, the personal computer followed a similar path of development. There was an enabling technology, the microprocessor, which took several years to be used in a product that the mass consumer wanted. The computer amateur played an important but underappreciated role in this transformation, not least by being a consumer for the first software companies—whose role was analogous to that of the radio broadcasters. And the personal computer spawned a major industry—with entrants coming from both entrepreneurial start-ups and established computer firms such as IBM.

Microprocessors

The enabling technology for the personal computer, the microprocessor, was developed during 1969 to 1971 in the semiconductor firm Intel. (Like many of the later developments in computer history, the microprocessor was independently invented in more than one place—but Intel was undoubtedly the most important locus.) Intel was founded in 1968 by Robert Noyce and Gordon Moore, both vice presidents of Fairchild Semiconductor and two of the original Shockley Eight. Today, Intel has annual revenues of several billion dollars and Noyce and Moore are legends of the American electronics industry. The microprocessor itself, however, was suggested not by them but by an Intel engineer Ted Hoff, then in his early thirties.

When Intel first began operations in 1968, it specialized in the manufacture of semiconductor memory and custom-designed chips. Intel's custom-chip sets were typically used in calculators, video games, electronic test gear, and control equipment. In 1969 Intel was approached by the Japanese calculator manufacturer Busicom to develop a chip set for a new scientific calculator—a fairly up-market model that would include trigonometrical and other advanced mathematical functions. The job of designing the chip set was assigned to Ted Hoff and his coworkers.

Hoff decided that instead of specially designed logic chips for the calculator, a better approach would be to design a general-purpose chip that could be programmed with the specific calculator functions. Such a chip would of course be a rudimentary computer in its own right, although it was some time before the significance of this dawned inside Intel.

The new calculator chip, known as the 4004, was delivered to Busicom in early 1971. Unfortunately, Busicom soon found itself a victim of the calculator price wars of the early 1970s and went into receivership. Before it did so, however, it negotiated the price of the 4004 downward in exchange for Intel acquiring the rights to market the new chip on its own account. Intel did this in November 1971, placing an advertisement in *Electronics News* that read: "Announcing a new era of integrated electronics: A microprogrammable computer on a chip." This first microprocessor sold for about $1,000.

The phrase "computer on a chip" was really copywriter's license; in any real application several other memory and controller chips would need to be attached to the 4004. But it was a potent metaphor that helped reshape the microelectronics industry over the next two years. During this period Intel replaced the 4004, a relatively low-powered device that processed only four bits of information at a time, with an eight-bit version, the 8008. A still more powerful chip, the 8080, which became the basis for several personal-computer designs, appeared in April 1974. By this time other semiconductor manufacturers were starting to produce their own microprocessors—such as the Motorola 6800, the Zilog Z80, and the MOS Technology 6502. With this competition, the price of microprocessors soon fell to around $100.

It would not be for another three years, however, that a real personal computer emerged, in the shape of the Apple II. The long gestation of the personal computer contradicts the received wisdom of its having arrived almost overnight. It was rather like the transition from wireless telegraphy to radio broadcasting, which the newspapers in 1921 saw as a "fad" that "seemed to come from nowhere"; in fact, it took several years, and the role of the hobbyist was crucial.

Computer Hobbyists and "Computer Liberation"

The computer hobbyist was typically a young male technophile. Most hobbyists had some professional competence. If not working with computers directly, they were often employed as technicians or engineers in the electronics industry. The typical hobbyist had cut his teeth in his early teens on electronic construction kits, bought through mail-order advertisements in one of the popular electronics magazines. Many of the hobbyists were active radio amateurs. But even those who were not radio amateurs owed much to the "ham" culture, which descended in an unbroken line from the early days of radio. After World War II, radio amateurs and electronic hobbyists moved on to building television sets and hi-fi kits advertised in magazines such as *Popular Electronics* and *Radio Electronics*. In the 1970s, the hobbyists lighted on the computer as the next electronic bandwagon.

Their enthusiasm for computing had often been produced by the hands-on experience of using a minicomputer at work or in college. The dedicated hobbyist hungered for a computer at home for recreational use, so that he could explore its inner complexity, experiment with computer games, and hook it up to other electronic gadgets. However, the cost of a minicomputer—typically $20,000 for a complete installation—was way beyond the pocket of the average hobbyist. To the nonhobbyist, why anyone would have wanted his own computer was a mystery: It was sheer techno-enthusiasm, and one can no more explain it than one can explain why people wanted to build radio sets sixty years earlier when there were no broadcasting stations.

It is important to understand that the hobbyist could conceive of hobby computing only in terms of the technology with which he was familiar. This was not the personal computer as we know it today; rather, the computing that the hobbyist had in mind in the early 1970s was a minicomputer hooked up to a teletype equipped with a paper-tape reader and punch for getting programs and data in and out of the machine. While teletypes were readily available in government-surplus shops, the most expensive part of the minicomputer—the central processing unit—remained much too costly for the amateur. The allure of the microprocessor was that it would reduce the price of the central processor by vastly reducing the chip count in the conventional computer.

The amateur computer culture was widespread. While it was particularly strong in Silicon Valley and around Route 128, computer hobbyists were to be found all over the country. The computer hobbyist was primarily interested in tinkering with computer hardware; software and applications were very much secondary issues.

Fortunately, the somewhat technologically fixated vision of the computer hobbyists was leavened by a second group of actors: the advocates of "computer liberation." It would, perhaps, be overstating the case to describe computer liberation as a movement, but there was unquestionably a widely held desire to bring computing to ordinary people. Computer liberation was particularly strong in California, and this perhaps explains why the personal computer was developed in California rather than (say) around Route 128.

Computer liberation sprang from a general malaise in the under-thirty crowd in the post-Beatles, post–Vietnam War period of the early 1970s. There was still a strong anti-establishment culture that expressed itself through the phenomena of college dropouts and campus riots, communal living, hippie culture, and alternative lifestyles sometimes associated with drugs. Such a movement for liberation would typically want to wrest communications technologies from vested corporate interests. In an earlier generation the liberators might have wanted to appropriate the press, but in fact the technology of printing and distribution channels were freely available, so that the young, liberal-minded community was readily able to communicate through magazines such as *Rolling Stone* as well as a vast underground press. On the other hand, computer technology was unquestionably not freely available; it was mostly rigidly controlled in government bureaucracies or private corporations. The much vaunted computer utility was, at $10 to $20 per hour, beyond the reach of ordinary users.

The most articulate spokesperson for the computer-liberation idea was Ted Nelson, the financially independent son of the Hollywood actress Celeste Holm. Among Nelson's radical visions of computing was an idea called *hypertext*, which he first described in the mid-1960s. Hypertext was a system by which an untrained person could navigate through a universe of information held on computers. Before such an idea could become a reality, however, it was necessary to "liberate" computing: to make it accessible to ordinary people at a trivial cost. In the 1970s Nelson promoted computer liberation as a regular speaker at computer hobbyist gatherings. He took the idea further in his self-published books *Computer Lib* and *Dream Machines,* which appeared in 1974. While Nelson's uncompromising views and his unwillingness to publish his books through conventional channels perhaps added to his anti-establishment appeal, this created a barrier between himself and the academic and commercial establishments.

It is not possible at this distance in time to properly evaluate Nelson's impact on the development of personal computing. In many ways his influence was similar, though certainly less important, to that of Marshall McLuhan, whose book *Understanding Media* provided such resonant late-twentieth-century ideas as "the medium is the message" and "the global village." In both cases their influence has been largely intangible, but it seems likely that cultural historians one day will see them as having changed the intellectual climate—in the case of McLuhan, who was well integrated in the academic and cultural establishment, on a global scale, while Nelson influenced mainly the young, predominantly male, local Californian technical community.

Personal computing in 1974, whether it was the vision of computer liberation or that of the computer hobbyist, bore little resemblance to the personal computer that emerged three years later—that is, the configuration of a self-contained machine, somewhat like a typewriter, with a keyboard and screen, an internal microprocessor-based computing engine, and a floppy disk for long-term data storage. In 1974 the computer-liberation vision of personal computing was a terminal attached to a large, information-rich computer utility at very low cost, while the computer hobbyist's vision was that of a traditional minicomputer. What brought together these two groups, with such different perspectives, was the arrival of the first hobby computer, the Altair 8800.

The Altair 8800 and Bill Gates

In January 1975 the first microprocessor-based computer, the Altair 8800, was announced on the front cover of *Popular Electronics.* The Altair 8800 is often described as the first personal computer. This was true only in the sense that its price was so low that it could be realistically bought by an individual. In every other sense the Altair 8800 was a traditional minicomputer. Indeed, the blurb on the front cover of *Popular Electronics* described it as exactly that: "Exclusive!

Altair 8800. The most powerful minicomputer project ever presented—can be built for under $400."

The Altair 8800 closely followed the marketing model of the electronic hobbyist kit: It was inexpensive ($397) and was sold by mail order as a kit that the enthusiast had to assemble himself. In the tradition of the electronics hobbyist kit, the Altair 8800 often did not work when the enthusiast had constructed it; and even if it did work, it did not do anything very useful. The computer consisted of a single box containing the central processor, with a panel of switches and lights on the front; it had no display, no keyboard, and not enough memory to do anything useful. Moreover, there was no way to attach a device such as a teletype to the machine to turn it into a useful computer system.

The only way the Altair 8800 could be programmed was by entering programs in pure binary code by flicking the hand switches on the front. When loaded, the program would run; but the only evidence of its execution was the change in the shifting pattern of the lights on the front. This limited the Altair 8800 to programs that only a dedicated computer hobbyist would ever be able to appreciate. Entering the program was extraordinarily tedious, taking several minutes—but as there were only 256 bytes of memory, there was a limit to the complexity of programs that could be attempted.

The Altair 8800 was produced by a tiny Albuquerque, New Mexico, electronics kit supplier, Micro Instrumentation Telemetry Systems (MITS). The firm had originally been set up by an electronics hobbyist, Ed Roberts, to produce radio kits for model airplanes. In the early 1970s Roberts began to sell kits for building electronic calculators, but that market dried up in 1974 during the calculator wars. Although he had toyed with the idea of a general-purpose computer for some time, it was only when the more obvious calculator market faded away that he decided to take the gamble.

The Altair 8800 was unprecedented and in no sense a "rational" product—it would appeal only to an electronics hobbyist of the most dedicated kind, and even that was not guaranteed. Despite its many shortcomings, the Altair 8800 was the grit around which the pearl of the personal-computer industry grew during the next two years. The limitations of the Altair 8800 created the opportunity for small-time entrepreneurs to develop "add-on" boards so that extra memory, conventional teletypes, and audiocassette recorders (for permanent data storage) could be added to the basic machine. Almost all of these start-up companies consisted of two or three people—mostly computer hobbyists hoping to turn their pastime to profit. A few other entrepreneurs developed software for the Altair 8800.

In retrospect, the most important of the early software entrepreneurs was Bill Gates, the co-founder of Microsoft. Although his ultimate financial success has been almost without parallel, his background was quite typical of a 1970s software nerd—a term that conjures up an image of a pale, male adolescent, lacking in social skills, programming by night and sleeping by day, oblivious to the wider world and the need to gain qualifications and build a career. This stereotype, though exaggerated, contains an essential truth; nor was it a new phenomenon—the programmer-by-night has existed since the 1950s. Indeed, programming the first personal computers had many similarities to programming a 1950s mainframe: There were no advanced software tools, and programs had to be hand-crafted in the machine's own binary codes so that every byte of the tiny memory could be used to its best advantage.

Gates, born in 1955 in Seattle to upper-middle-class parents, was first exposed to computers in 1969, when he learned to program in BASIC using a commercial time-sharing system on which his high school rented time. He and his close friend, Paul Allen, two years his senior, discovered a mutual passion for programming. They also shared a strong entrepreneurial flair from the very beginning: When Gates was only sixteen, long before the personal-computer revolution, the two organized a small firm for the computer analysis of traffic data, which they named Traf-O-Data. While Allen went on to study computer science at Washington State University, Gates decided—under the influence of his lawyer father—to prepare for a legal career at Harvard University, where he enrolled in the fall of 1973. However, he soon found that his studies did not engage his interest, and he continued to program by night.

The launch of the Altair 8800 in 1975 transformed Gates's and Allen's lives. Almost as soon as they heard of the machine, they recognized the software opportunity it represented and proposed to MITS' Ed Roberts that they should develop a BASIC programming system for the new machine. Besides being easy to develop, BASIC was the language favored by the commercial time-sharing systems and minicomputers that most computer hobbyists had encountered, and would therefore be the ideal vehicle for the personal-computer market. Roberts was enthusiastic, not least because BASIC would need a lot more memory to run than was normally provided with the Altair 8800; he expected to be able to sell extra memory with a high margin of profit.

Gates and Allen formed a partnership they named Micro-Soft (the hyphen was later dropped), and after six weeks of intense programming effort they delivered a BASIC programming system to MITS in February 1975. Now graduated, Allen became software director at MITS—a somewhat overblown job title for what was still a tiny firm located in a retail park. Gates remained at Harvard for a few more months, more from inertia than vocation; by the end of the academic year the direction of the booming microcomputing business was clear, and Gates abandoned his formal education. During the next two years, literally hundreds of small firms entered the microcomputer software business, and Microsoft was by no means the most prominent.

The Altair 8800, and the add-on boards and software that were soon available for it, transformed hobby electronics in a way not seen since the heyday of radio. In the spring of 1975, for example, the "Homebrew Computer Club" was established in Menlo Park, on the edge of Silicon Valley. Besides acting as a swap shop for computer components and programming tips, it also provided a forum for the computer-hobbyist and computer-liberation cultures to meld.

During the first quarter of 1975, MITS received orders worth over $1 million for the Altair 8800 and launched its first "world-wide" conference. Speakers at the conference included Ed Roberts, Gates and Allen as the developers of Altair BASIC, and the computer-liberation guru Ted Nelson. At the meeting Gates launched a personal diatribe against hobbyists who pirated software. This was a dramatic position: He was advocating a shift in culture from the friendly sharing of free software among hobbyists to that of an embryonic branch of the packaged-software industry. Gates encountered immense hostility—his speech was, after all, the very antithesis of computer liberation. But his position was eventually accepted by producers and consumers, and over the next two years it was instrumental in transforming the personal computer from a utopian ideal to an economic artifact.

The period 1975–77 was a dramatic and fast-moving one in which the microcomputer was transformed from a hobby machine to a consumer product. The outpouring of newly launched computer magazines remains the most permanent record of this frenzy. Some of them, such as *Byte* and *Popular Computing*, followed in the tradition of the electronics hobby magazines, while others, such as the whimsically titled *Dr. Dobb's Journal of Computer Calisthenics and Orthodontia*, responded more to the computer-liberation culture. The magazines were important vehicles for selling computers by mail order, in the tradition of hobby electronics. Mail order was soon supplanted, however, by computer shops such as the Byte Shop and ComputerLand, which initially had the ambiance of an electronics hobby shop: full of dusty, government-surplus hardware, and electronic gadgets. Within two years, ComputerLand would be transformed into a nationwide chain, stocking shrink-wrapped software and computers in colorful boxes.

While it had taken the mainframe a decade to be transformed from laboratory instrument to business machine, the personal computer was transformed in just two years. The reason for this rapid development was that most of the subsystems required to create a personal computer already existed: keyboards, screens, disk drives, and printers. It was just a matter of putting the pieces together. Hundreds of firms—not just on the West Coast, but all over the country—sprang up over this two-year period. They were mostly tiny start-ups, consisting of a few computer hobbyists or young computer professionals; they supplied complete computers, add-on boards, peripherals, or software. Within months of its initial launch at the beginning of 1975, the Altair 8800 had itself been eclipsed by dozens of new models produced by firms such as Applied Computer Technology, IMSAI, North Star, Cromemco, and Vector.

The Rise of Apple Computer

Most of the new computer firms fell almost as quickly as they rose, and only a few survived beyond the mid-1980s. Apple Computer was the rare exception in that it made it into the Fortune 500 and achieved long-term global success. Its initial trajectory, however, was quite typical of the early hobbyist start-ups.

Apple was founded by two young computer hobbyists, Stephen Wozniak and Steve Jobs. Wozniak grew up in Cupertino, California, in the heart of the booming West Coast electronics industry. Like many of the children in the area, electronics was in the air they breathed. Wozniak took to electronics almost as soon as he could think abstractly; he was a talented hands-on engineer, lacking any desire for a deeper, academic understanding. He obtained a radio amateur operating license while in sixth grade, graduated to digital electronics as soon as integrated circuits became available in the mid-1960s, and achieved a little local celebrity by winning an interschools science prize with the design of a simple adding circuit. Unmotivated by academic studies, he drifted in and out of college without gaining significant qualifications, although he gained a good working knowledge of minicomputers.

Like many electronics hobbyists, Wozniak dreamed of owning his own minicomputer, and in 1971 he and a friend went so far as to construct a rudimentary machine from parts rejected by local companies. It was around this time that he teamed up with Steve Jobs, five years his junior, and together they went into business making "blue boxes"—gadgets that mimicked dial tones, enabling telephone calls to be made for free. While not illegal to make and sell, using a blue box was illegal, as it defrauded the phone companies of revenues; but many of these hobbyists regarded it as a victimless crime, and in the moral climate of the West Coast computer hobbyist it was pretty much on a par with pirating software. This in itself is revealing of how far cultural attitudes would shift as the personal computer made the transition from hobby to industry.

Despite his lack of formal qualifications, Wozniak's engineering talent was recognized and he found employment in the calculator division of Hewlett-Packard in 1973; but for what amounted to a late-twentieth-century form of patronage that prevailed in the California electronics industry, Wozniak might have found his career confined to that of a low-grade technician or repairman.

While Wozniak was a typical, if unusually gifted, hobbyist, Steve Jobs bridged the cultural divide between computer hobbyism and computer liberation. That Apple Computer ultimately became a global player in the computer industry is largely due to Jobs's evangelizing of the personal computer, his ability to harness Wozniak's engineering talent, and his willingness to seek out the organizational capabilities needed to build a business.

Born in 1955, Jobs was brought up by adoptive blue-collar parents. Although not a child of the professional electronics engineering classes, Jobs took to the electronic hobbyism that he saw all around him. While a capable enough engineer, he was not in the same league as Wozniak. There are many stories of Jobs's astounding, and sometimes overbearing, self-confidence, which had a charm when he was young but was seen as autocratic and immature when he became the head of a major corporation. One of the more celebrated stories is that, at the age of thirteen, when he needed some electronic components for a school project, he telephoned William Hewlett, the multimillionaire co-founder of Hewlett-Packard. Hewlett, won over by Jobs's chutzpah, not only gave him the parts but offered him a part-time job with the company.

Something of a loner, and not academically motivated, Jobs drifted in and out of college in the early 1970s before finding a well-paid niche as a games designer for Atari. An admirer of the Beatles, like them Jobs spent a year pursuing transcendental meditation in India and turned vegetarian. Jobs and Wozniak made a startling contrast: Wozniak was the archetypal electronic hobbyist with social skills to match, while Jobs affected an aura of inner wisdom, wore open-toed sandals, had long, lank hair, and sported a Ho Chi Minh beard.

The turning point for both Jobs and Wozniak was attending the Homebrew Computer Club in early 1975. Although Wozniak knew about microprocessors from his familiarity with the calculator industry, he had not up to that point realized that they could be used to build general-purpose computers and had not heard of the Altair 8800. But he had actually built a computer, which was more than could be said of most Homebrew members at that date, and he found himself among an appreciative audience. He quickly took up the new microprocessor technology, and within a few weeks had thrown together a computer based on the MOS Technology 6502 chip. He and Jobs called it the "Apple," for reasons that are now lost in time, but possibly for the Beatles' record label.

While Jobs never cared for the "nit-picking technical debates" of the Homebrew computer enthusiasts, he did recognize the latent market they represented. He therefore cajoled Wozniak into developing the Apple computer and marketing it, initially through the Byte Shop. The Apple was a very crude machine, consisting basically of a naked circuit board, lacking a case, a keyboard, or screen, or even a power supply. Eventually about two hundred were sold, each hand-assembled by Jobs and Wozniak in the garage of Jobs's parents.

In 1976 Apple was just one of dozens of computer firms competing for the dollars of the computer hobbyist. Jobs recognized before most, however, that the microcomputer had the potential to be a consumer product for a much broader market if it were appropriately packaged. To be a success as a product, the microcomputer would have to be presented as a self-contained unit in a plastic case, able to be plugged into a standard household outlet just like any other appliance; it would need a keyboard to enter data, a screen to view the results of a computation, and some form of long-term storage to hold data and programs. Most important, the machine would need software to appeal to anyone other than an enthusiast. First this would be BASIC, but eventually a much wider range of software would be required. This, in a nutshell, was the specification for the Apple II that Jobs passed down to Wozniak to create.

For all his naïveté as an entrepreneur Jobs understood, where few of his contemporaries did, that if Apple was to become a successful company, it would need access to capital, professional management, public relations, and distribution channels. None of these was easy to find at a time when the personal computer was unknown outside hobbyist circles. Jobs's evangelizing was called on in full measure to acquire these capabilities. During 1976, while Wozniak designed the Apple II, Jobs secured venture capital from Mike Markkula, to whom he had been introduced by his former employer at Atari, Nolan Bushnell. Markkula was a thirty-four-year-old former Intel executive who had become independently wealthy from stock options. Through Markkula's contacts, Jobs located an experienced young professional manager from the semiconductor industry, Mike Scott, who agreed to serve as president of the company. Scott would take care of operational management, leaving Jobs free to evangelize and determine the strategic direction of Apple. The last piece of Jobs's plan fell into place when he persuaded the prominent public relations company Regis McKenna to take on Apple as a client.

Throughout 1976 and early 1977, while the Apple II was perfected, Apple Computer remained a tiny company with fewer than a dozen employees occupying 2,000 square feet of space in Cupertino, California.

VisiCalc

During 1977 three distinct paradigms for the personal computer emerged, represented by three leading manufacturers: Apple, Commodore Business Machines, and Tandy, each of which defined the personal computer in terms of its own existing culture and corporate outlooks.

If there can be said to be a single moment when the personal computer arrived in the public consciousness, then it was at the West Coast Computer Faire in April 1977, when the first two machines for the mass consumer, the Apple II and the Commodore PET, were launched. Both machines were instant hits, and for a while they vied for market leadership. At first glance the Commodore PET looked very much like the Apple II in that it was a self-contained appliance with a keyboard, a screen, a means of program storage, and with BASIC ready-loaded so that users could write programs.

The Commodore PET, however, coming from Commodore Business Machines—a firm that had originally made electronic calculators—was not so much a computer as a calculator writ large. For example, the keyboard had the tiny buttons of a calculator keypad rather than the keyboard of a standard computer terminal. Moreover, like a calculator, the PET was a closed system, with no potential for add-ons such as printers or floppy disks. Nevertheless, this narrow specification and the machine's low price appealed to the educational market, where it found a niche supporting elementary computer studies and BASIC programming; eventually several hundred thousand machines were sold.

By contrast, the Apple II, although more expensive than the PET (it cost $1,298, excluding a screen), was a true computer system with the full potential for adding extra boards and peripherals. The Apple II was therefore far more appealing to the computer hobbyist because it offered the opportunity to engage with the machine by customizing it and using it for novel applications that the inventors could not envisage.

In August 1977 the third major computer vendor, Tandy, entered the market, when it announced its TRS-80 computer for $399. Produced by Tandy's subsidiary, Radio Shack, the TRS-80 was aimed at the retailer's existing customers, who consisted mainly of electronic hobbyists and buyers of video games. The low price was achieved by the user having to use a television set for a screen and an audiocassette recorder for program storage. The resulting hook-up was no hardship to the typical Tandy customer, although it would have been out of place in an office.

Thus, by the fall of 1977, although the personal computer had been defined physically as an artifact, a single constituency had not yet been established. For Commodore the personal computer was seen as a natural evolution of its existing calculator line. For Tandy it was an extension of its existing electronic-hobbyist and video games business. For Apple the machine was initially aimed at the computer hobbyist.

Jobs's ambition and vision went beyond the hobby market, and he envisioned the machine also being used as an appliance in the home—perhaps the result of his experience as a designer of domestic video games. This ambiguity was revealed by the official description of the Apple II as a "home/personal computer." The advertisement that Regis McKenna produced to launch the Apple II showed a housewife doing kitchen chores, while in the background her husband sat at the kitchen table hunched over an Apple II, seemingly managing the household's information. The copy read:

> The home computer that's ready to work, play and grow with you. . . . You'll be able to organize, index and store data on household finances, income taxes, recipes, your biorhythms, balance your checking account, even control your home environment.

These domestic projections for the personal computer were reminiscent of those for the computer utility in the 1960s, and were equally misguided. Moreover, the advertisement did not point out that these domestic applications were pure fantasy—there was no software available for "biorhythms," accounts, or anything else.

The constituency for the personal computer would be defined by the software that was eventually created for it.

At that time it was very easy to set up as a personal-computer software entrepreneur: All one needed was a machine on which to develop the software and the kind of programming know-how possessed by any talented first-year computer science student, which many hobbyists had already picked up in their teenage years. The barriers to entry into personal-computer software were so low that literally *thousands* of firms were established—and their mortality rate was phenomenal.

Up to 1976 there were only a handful of personal-computer software firms, mainly producing "system" software. The most popular products included Microsoft's BASIC programming language and Digital Research's CP/M operating system, which were each used in many different makes of computer. This software was usually bundled with the machine, and the firm was paid a royalty included in the overall price of the computer. In 1977 personal-computer software was still quite a small business: Microsoft had just five employees and annual sales of only $500,000.

With the arrival of consumer-oriented machines such as the Apple II, the Commodore PET, and the Tandy TRS–80, however, the market for "applications" software took off. Applications software enabled a computer to perform useful tasks without the owner having to program the machine directly. There were three main markets for applications software: games, education, and business.

The biggest market, initially, was for games software, which reflected the existing hobbyist customer base:

> When customers walked into computer stores in 1979, they saw racks of software, wall displays of software, and glass display cases of software. Most of it was games. Many of these were outer space games—*Space, Space II, Star Trek*. Many games appeared for the Apple, including Programma's simulation of a video game called *Apple Invaders*. Companies such as Muse, Sirius, Broderbund, and On-Line Systems reaped great profits from games.

Computer games are often overlooked in discussions of the personal-computer software industry, but they played an important role in its early development. Programming computer games created a corps of young programmers who were very sensitive to what we now call human-computer interaction. The most successful games were ones that needed no manuals and gave instant feedback. The most successful business software had similar, user-friendly characteristics. As for the games software companies themselves, the great majority of them faded away. While a handful of firms became major players, the market for recreational software never grew as large as that for business applications, and so none of the firms became Fortune 500 companies.

The second software market was for educational programs. Schools and colleges were the first organizations to buy personal computers on a large scale: Software was needed to learn mathematics; simulation programs were needed for science teaching; and programs were needed for business games, language learning, and music. Much of this early software was developed by teachers and students in their own time and was of rather poor quality. Some major programs were developed through research grants, but because of charitable status, the software was either free or sold on a nonprofit basis. As a result the market for educational software did not develop for a decade, and even today it is a poorly developed sector of personal-computer software.

The market of packaged software for business applications developed between 1978 and 1980, when three generic applications enabled the personal computer to become an effective business machine: the spreadsheet, the word processor, and the database. All these types of software already existed in the ordinary mainframe computer context, typically using a time-sharing terminal, so it was not obvious at the outset that the personal computer offered any advantage as a business machine.

The first application to receive wide acceptance was the VisiCalc spreadsheet. The originator of VisiCalc was a twenty-six-year-old Harvard MBA student, Daniel Bricklin, who thought of the idea of using a personal computer as a financial analysis tool, as an alternative to using a conventional mainframe computer or a time-sharing terminal. Bricklin sought the advice of a number of people, including his Harvard professor-supervisor, but they were somewhat discouraging because his idea seemed to offer no obvious advantage over a conventional computer. Bricklin was not dissuaded, however, and during 1977–78 he went into partnership with a programmer friend, Bob Frankson. In their spare time they developed a program for the Apple II computer. To market the program, Bricklin approached a former student from his MBA course, who was then running a company called Personal Software, which specialized in selling games software. They decided to call the program VisiCalc, for *Visi*ble *Calc*ulator.

Bricklin's program used about 25,000 bytes of memory, which was about as big as a personal computer of the period could hold, but was decidedly modest by mainframe standards. The personal computer, however, offered some significant advantages that were not obvious at the outset. Because the personal computer was a stand-alone, self-contained system, changes to a financial model were displayed almost instantaneously compared with the minute or so it would have taken on a conventional computer. This fast response enabled a manager to explore a financial model with great flexibility, asking what were known as "what if?" questions. It was almost like a computer game for executives.

When it was launched in December 1979, VisiCalc was a word-of-mouth success. Not only was the program a breakthrough as a financial tool but its users experienced for the first time the psychological freedom of having a machine of one's own, on one's desk, instead of having to accept the often mediocre take-it-or-leave-it services of a computer center. Moreover, at $3,000, including software, it was possible to buy an Apple II and VisiCalc out of a departmental, or even a personal, budget.

The success of VisiCalc has become one of the great heroic episodes of the personal-computer revolution and is often, alone, credited with transforming the industry. As described in Robert Slater's *Portraits in Silicon:*

> Suddenly it became obvious to businessmen that they had to have a personal computer: VisiCalc made it feasible to use one. No prior technical training was needed to use the spreadsheet program. Once, both hardware and software were for hobbyists, the personal computer a mysterious toy, used if anything for playing games. But after VisiCalc the computer was recognized as a crucial tool.

One can find a similar passage in virtually every history of the personal computer. On the whole, the role of VisiCalc has been exaggerated. Apple itself estimated that only 25,000 of the 130,000 computers it sold before September 1980 were bought on the strength of VisiCalc. Important as VisiCalc was, it seems highly likely that if it had not existed, then a word-processor or database application would have brought the personal computer into corporate use by the early 1980s.

Word processing on personal computers did not develop until about 1980. One reason for this was that the first generation of personal computers displayed only forty uppercase letters across the screen, and good-quality printers were expensive. This did not matter much when using a spreadsheet, but it made a personal computer much less attractive for word processing than an electric typewriter or a dedicated word-processing system. By 1980, however, new computers were coming onto the market capable of displaying eighty letters across the screen, including both upper and lower cases. The new computers could display text on the screen that was identical to the layout of the printed page—known as "what you see is what you get," or WYSIWYG. Previously, this facility had been available only in a top-of-the-line word processor costing several thousand dollars. The availability of low-cost printers that produced reasonable quality output, primarily from Japanese manufacturers, also greatly helped the word-processing market.

The first successful firm to produce word-processing software was MicroPro, founded by the entrepreneur Seymour Rubinstein in 1978. Rubinstein, then in his early forties, was formerly a mainframe software developer. He had a hobbyist interest in amateur radio and electronics, however, and when the first microcomputer kits became available, he bought one. He recognized very early on the personal computer's potential as a word processor, and he produced a program called WordMaster in 1978. This was replaced in mid-1979 with a full WYSIWYG system called WordStar, which quickly gained a two-thirds market share. WordStar sold hundreds of copies a month, at $450 a copy. During the next five years MicroPro sold nearly a million copies of its processing software and became a $100-million-a-year business.

During 1980, with dozens of spreadsheet and word-processing packages on the market and the launch of the first database products, the potential of the personal computer as an office machine became clearly recognizable. At this point the traditional business-machine manufacturers, such as IBM, began to take an interest.

The Reemergence of IBM

IBM was not, in fact, the giant that slept soundly during the personal-computer revolution. IBM had a sophisticated market research organization that attempted to predict market trends. The company was well aware of microprocessors and personal computers. Indeed, in 1975 it had developed a desktop computer for the scientific market (the model 5100), but it did not sell well. By 1980 IBM was selling a dedicated word processor based on microprocessor technology. But its sales came a poor second to its traditional electric typewriters, of which IBM was still selling a million a year.

Once the personal computer became clearly defined as a business machine in 1980, IBM reacted with surprising speed. The proposal that IBM should enter the personal-computer business came from William C. Lowe, a senior manager who headed the company's "entry-level systems" division in Boca Raton, Florida. In July 1980 Lowe made a presentation to IBM's senior management in Armonk, New York, with a radical plan: Not only should IBM enter the personal-computer market but it should also abandon its traditional development processes in order to match the dynamism of the booming personal-computer industry.

For nearly a century IBM had operated a bureaucratic development process by which it typically took three years for a new product to reach the market. Part of the delay was due to IBM's century-old vertical integration practice, by which it maximized profits by manufacturing in-house all the components used in its products: semiconductors, switches, plastic cases, and so on. Lowe argued that IBM should instead adopt the practice of the rest of the industry by outsourcing all the components it did not already have in production, including software. Lowe proposed yet another break with tradition—that IBM should not use its direct sales force to sell the personal computer but should instead use regular retail channels.

Surprisingly, in light of its stuffy image, IBM's top management agreed to all that Lowe recommended, and within two weeks of his presentation he was authorized to go ahead and build a prototype, which had to be ready for the market within twelve months. The development of the personal computer would be known internally as Project Chess.

IBM's relatively late entry into the personal-computer market gave it some significant advantages. First, it could make use of the second generation of microprocessors (which processed sixteen bits of data at a time instead of eight); this would make the IBM personal computer significantly faster than any other machine on the market. IBM chose to use the Intel 8088 chip, thereby guaranteeing Intel's future prosperity.

Although IBM was the world's largest software developer, paradoxically it did not have the skills to develop software for personal computers. Its bureaucratic software development procedures were slow and methodical, and geared to large software artifacts; the company lacked the critical skills needed to develop the "quick-and-dirty" software needed for personal computers.

IBM initially approached Gary Kildall of Digital Research—the developer of the CP/M operating system—for operating software for the new computer, and herein lies one of the more poignant stories in the history of the personal computer. For reasons now muddied, Kildall blew the opportunity. One version of the story has it that he refused to sign IBM's nondisclosure agreement, while another version has him doing some recreational flying while the dark-suited IBMers cooled their heels below. In any event, the opportunity passed Digital Research by and moved on to Microsoft. Over the next decade, buoyed by the revenues from its operating system for the IBM personal computer, Microsoft became the quintessential business success story of the late twentieth century, and Gates became a billionaire at the age of thirty-one. Hence, for all of Gates's self-confidence and remarkable business acumen, he owed almost everything to being in the right place at the right time.

The IBM entourage arrived at Bill Gates and Paul Allen's Microsoft headquarters in July 1980. It was then a tiny (thirty-two-person) company located in rented offices in downtown Seattle. It is said that Gates and Allen were so keen to win the IBM contract that they actually wore business suits and ties. Although Gates may have appeared a somewhat nerdish twenty-five-year-old who looked fifteen, he came from an impeccable background, was palpably serious, and showed a positive eagerness to accommodate the IBM culture. For IBM, he represented as low a risk as any of the personal-computer software firms, almost all of which were noted for their studied contempt for Big Blue. It is said that when John Opel, IBM's president, heard about the Microsoft deal, he said, "Is he Mary Gates's son?" He was. Opel and Gates's mother both served on the board of the United Way.

At the time that Microsoft made its agreement with IBM for an operating system, it did not have an actual product, nor did it have the resources to develop one in IBM's time scale. However, Gates obtained a suitable piece of software from a local software firm, Seattle Computer Products, for $30,000 cash and improved it. Eventually, the operating system, known as MS-DOS, would be bundled with almost every IBM personal computer and compatible machine, earning Microsoft a royalty of between $10 and $50 on every copy sold.

By the fall of 1980 the prototype personal computer, known internally as the Acorn, was complete; IBM's top management gave final authorization to go into production. Up to this point the Acorn had been only a development project like any other—now serious money was involved. Lowe, his mission essentially accomplished, moved up into the higher echelons of IBM, leaving his second-in-command, Don Estridge, in overall charge. Estridge was an unassuming forty-two-year-old. Although, as the corporate spokesman for the IBM personal computer, he later became as well known as any IBMer apart from the company's president, he never attracted as much media attention as the Young Turks such as Gates and Jobs.

The development team under Estridge was now increased to more than a hundred, and factory arrangements were made for IBM to assemble computers using largely outsourced components. Contracts for the bulk supply of subsystems were finalized with Intel for the 8088 microprocessor, with Tandon for floppy disk drives, with Zenith for power supplies, and with the Japanese company Epson for printers. Contracts were also firmed up for software. Besides Microsoft for its operating system and BASIC, arrangements were made to develop a version of the VisiCalc spreadsheet, a word processor, and a suite of business programs. A games program, Adventure, was also included with the machine, suggesting that even at this late date it was not absolutely clear whether the personal computer was a domestic machine, a business machine, or both.

Not everyone in IBM was happy to see the personal computer—whether for home or business—in the company's product line. One insider was reported as saying:

> Why on earth would you care about the personal computer? It has nothing at all to do with office automation. It isn't a product for big companies that use "real" computers. Besides, nothing much may come of this and all it can do is cause embarrassment to IBM, because, in my opinion, we don't belong in the personal computer business to begin with.

Overriding these pockets of resistance inside the company, IBM began to actively consider marketing. The economics of the personal computer determined that it could not be sold by IBM's direct sales force because the profit margins would be too slender. The company negotiated with the Chicago-based Sears Company to sell the machine at its Business Centers and contracted with ComputerLand to retail the machine in its stores. For its traditional business customers, IBM would also sell the machines in its regular sales offices, alongside office products such as electric typewriters and word processors.

Early in 1981, only six months after the inception of Project Chess, IBM appointed the West Coast–based Chiat Day advertising agency to develop an advertising campaign. Market research suggested that the personal computer still lay in the gray area between regular business equipment and a home machine. The advertising campaign was therefore ambiguously aimed at both the business and home user. The machine was astutely named the IBM Personal Computer, suggesting that the IBM machine and the personal computer were synonymous. For the business user, the fact that the machine bore the IBM logo was sufficient to legitimate it inside the corporation. For the home user, however, market research revealed that although the personal computer was perceived as a good thing, it was also seen as intimidating—and IBM itself was seen as "cold and aloof." The Chiat Day campaign attempted to allay these fears by featuring in its advertisements a Charlie Chaplin lookalike and alluding to Chaplin's famous movie *Modern Times*. Set in a futuristic automated factory, *Modern Times* showed the "little man" caught up in a world of hostile technology, confronting it, and eventually overcoming it. The Charlie Chaplin figure reduced the intimidation factor and gave IBM "a human face."

During the summer of 1981 the first machines began to come off the IBM assembly plant in Boca Raton, and by early August initial shipments totaling 1,700 machines had been delivered to Sears Business Centers and ComputerLand stores ready for the launch. A fully equipped IBM Personal Computer, with 64 Kbytes of memory and a floppy disk, cost $2,880.

The IBM Personal Computer was given its press launch in New York on 12 August. There was intense media interest, which generated many headlines in the computer and business press. In the next few weeks the IBM Personal Computer became a runaway success that exceeded almost everyone's expectations, inside and outside the company. While many business users had hesitated over whether to buy an Apple or a Commodore or a Tandy machine, the presence of the IBM logo convinced them that the technology was for real: IBM had legitimated the personal computer. There was such a demand for the machine that production could not keep pace, and retailers could do no more than placate their customers by placing their names on a waiting list. Within days of the launch, IBM decided to quadruple production.

During 1982–83 the IBM Personal Computer became an industry standard. Most of the popular software packages were converted to run on the machine, and the existence of this software reinforced its popularity. This encouraged other manufacturers to produce "clone" machines, which ran the same software. This was very easy to do because the Intel 8088 microprocessor used by IBM and almost all the other subsystems were readily available on the open market. Among the most successful of the clone manufacturers was Houston-based Compaq, which produced its first machine in 1982. In its first full year of business, it achieved sales of $110 million. Adroitly swimming with the tide, several of the leading manufacturers such as Tandy, Commodore, Victor, and Zenith switched into making IBM-compatible products. Alongside the clone manufacturers, a huge subindustry developed to manufacture peripherals, memory boards, and add-ons. The software industry published thousands of programs for the IBM-compatible personal computer—or the IBM PC, as the machine soon became known. In 1983 it was estimated that there were a dozen monthly magazines and a score of weekly newspapers for users of the machine. Most famously, in January 1983, the editors of *Time* magazine nominated as their Man of the Year not a person but a machine: the PC.

Almost all the companies that resisted the switch to the IBM standard soon went out of existence or were belatedly forced into conforming. The only important exception was Apple Computer, whose founder, Steve Jobs, had seen another way to compete with the IBM standard: not by making cheaper hardware but by making better software.

11

:: BROADENING THE APPEAL

MOST NEW technologies start out imperfect, difficult to master, and capable of being used only by enthusiastic early adopters. In the early 1920s, radios used crude electronic circuits that howled and drifted and needed three hands to tune; they ran on accumulators that had to be taken to the local garage for recharging, and the entertainment available often consisted of nothing more than a broadcast from the local dance hall. But over a period of a few years, the superhetrodyne circuit transformed the stability and ease of tuning of radio sets, and it became possible to plug one into any electrical outlet; the quality of programs—drama, music, news, and sports—came to rival that available in the movie theaters. Soon "listening in" was an everyday experience.

Something very similar happened to personal computers in the 1980s, so that people of ordinary skill would be able to use them and want to use them. The graphical user interface made computers much easier to use, while software and services made them worth owning. A new branch of the software industry created thousands of application programs, and the CD-ROM brought information in book-like quantities to the desktop. And when computer networks enabled users to reach out to other individuals, by means of "chat rooms" or exchanging e-mail, the personal computer had truly become an information machine.

The Maturing of the Personal-Computer Software Industry

On 24 August 1995 Microsoft launched Windows 95, its most important software product to date. The advance publicity was unparalleled. In the weeks

231

before the launch, technology stocks were driven sharply higher in stock markets around the world, and in the second half of August the Microsoft publicity juggernaut reached maximum speed. Press reports estimated that the Windows 95 launch cost as much as $200 million, of which $8 million alone was spent acquiring the rights to use the Rolling Stones hit "Start Me Up" as background music for the TV commercials. In major cities theaters were hired and video screens installed so that Microsoft's chairman, Bill Gates—said to be the richest man in the world, with personal assets of $12 billion—could deliver his address to the waiting world.

When we left Microsoft in the last chapter, in 1980, it was a tiny outfit with thirty-eight employees and annual sales of just $8 million. Ten years later, in 1990, it had 5,600 employees and sales of $1.8 billion. As suggested by the rise of Microsoft, the important personal-computer story of the 1980s was not hardware but software.

The personal-computer software industry developed in two phases. The first phase, which can be characterized as the gold-rush era, lasted from about 1975 to 1982; during this period, barriers to entry were extremely low, and there were several thousand new entrants, almost all of which were undercapitalized two- or three-person start-ups. The second phase, which began about 1983, following the standardization of the personal-computer market around the IBM-compatible PC, was a period of consolidation in which many of the early firms were shaken out, new entrants required heavy inputs of venture capital, and a small number of (mainly American) firms emerged as global players.

Apart from its meteoric growth, the most remarkable aspect of the industry was its almost total disconnection from the existing packaged-software industry, which in 1975 was a billion-dollar-a-year business with several major international suppliers. There were both technical and cultural reasons for this failure to connect. The technical reason was that the capabilities of the existing software firms—with their powerful software tools and methodologies for developing large, reliable programs—were irrelevant for developing programs for the tiny memories of the first personal computers; indeed, they were likely to be counterproductive. Entrants to the new industry needed not advanced software-engineering knowledge but the same kind of savvy as the first software contractors in the 1950s: creative flair and the technical knowledge of a bright undergraduate. The existing software companies simply could not think or act small enough; their overhead costs did not allow them to be cost-competitive with software products for personal computers.

But the cultural reasons were at least as important. Whereas the traditional packaged-software firms marketed their programs using dark-suited salespeople with IBM-type backgrounds, the personal-computer software companies sold their products through mail order and retail outlets. And if anyone attended an industry gathering wearing a tie, someone would "cut it off and throw you in the pool."

Despite the large number of entrants in the industry, a small number of products quickly emerged as market leaders. These included the VisiCalc spreadsheet, the WordStar word processor, and the dBase database. By the end of 1983 these three products dominated their respective markets, with cumulative sales of 800,000, 700,000, and 150,000, respectively.

Personal-computer software was a new type of product that had to evolve its own styles of marketing. When industry pundits searched for an analogy, they often likened the personal-computer software business to pop music or book publishing. For example, a critical success factor was marketing. Advertising costs typically took up 35 cents of each retail dollar. Promotions took the form of magazine advertising, free demonstration diskettes, point-of-sale materials, exhibitions, and so on. Marketing typically accounted for twice the cost of actually developing a program. As one industry expert put it, the barriers to entry into the software business were "marketing, marketing, and marketing." By contrast, the manufacturing cost—which consisted simply of duplicating floppy disks and manuals—was the smallest cost component of all.

The analogy to pop music and book publishing was apt. Every software developer was seeking that elusive "hit," so that the marketing and R&D costs would be spread over as high a number of sales as possible.

By about 1983 the gold rush was over. It was estimated that fifteen companies had two-thirds of the market, and three significant barriers had been erected to entry into the personal-computer software business. The first was technological, caused by the dramatically improving performance of personal computers. The new generation of IBM-compatible PCs that had begun to dominate the market were capable of running software comparable with that used on small mainframes and required similar technological resources for its development. Whereas in 1979 major software packages had been written by two or three people, now teams of ten and often many more were needed. (To take one example, while the original VisiCalc had contained about 10,000 instructions, mature versions of the Lotus 1-2-3 spreadsheet contained about 400,000 lines of code.) The second barrier to entry was know-how. The sources of knowledge of how to create personal-computer software with an attractive interface had become locked into the existing firms. This knowledge was not something that could be learned from the literature or in a computer science class. The third, and probably the greatest, barrier was access to distribution channels. In 1983 it was said that there were 35,000 products competing for a place among the 200 products that a typical computer store could stock—there were 300 word-processing packages just for the IBM-compatible PC. A huge advertising expenditure—and therefore a large injection of capital—was needed to overcome this barrier.

One might have expected that these barriers to entry would have protected the existing firms such as VisiCorp, MicroPro, and Ashton-Tate, but this was not the case. By 1990 these and most of the other prominent software firms of the early 1980s had become also-rans, been taken over, or gone out of business altogether.

The reasons for this transformation are complex, but the dominating cause was the importance of a single "hit" product whose arrival could transform a balance sheet for several years, but whose demise could send the firm into a spiral of decline. Perhaps the most poignant of these dramatic turns of fortune was that of VisiCorp, the publisher of VisiCalc. At its peak in 1983, VisiCorp had annual revenues of $40 million, but by 1985 it had ceased to exist as an independent entity. VisiCorp was effectively wiped out by the arrival of a competing product, Lotus 1-2-3.

To create a software hit, one needed access either to big sources of venture capital or to a healthy revenue stream from an existing successful product. The story of the Lotus Development Corporation, which was formed by a thirty-two-year-old entrepreneur, Mitch Kapor, in 1982, illustrates the financial barriers that had to be overcome to establish a successful start-up in computer software. Kapor was a charismatic freelance software developer who had produced a couple of successful packages in 1979 for VisiCorp. Kapor had originally received a royalty on sales from VisiCorp, but he decided to sell all the rights to his packages to the company for a single payment of $1.7 million. He used this cash—plus further venture-capital funds totaling $3 million—to develop a new spreadsheet called Lotus 1-2-3, which would compete head-on with the top-selling package, VisiCalc. To beat VisiCalc in the marketplace, Lotus 1-2-3 needed to be a much more technically sophisticated package, and estimates were that it cost $1 million to develop. It was reported that Lotus spent a further $2.5 million on the initial launch of its new spreadsheet. Of the retail price of $495, it was said that about 40 percent went into advertising. With this blaze of publicity, Lotus 1-2-3 sold 850,000 copies in its first eighteen months and instantly became the market leader.

The story of Microsoft illustrates the value of an income stream from an already successful product as an alternative to raising venture capital. By 1990

Microsoft had emerged as the outstanding leader of the personal-computer software industry, and biographies and profiles of its founder, William Henry Gates III, poured forth. In 1981, when firms such as MicroPro and VisiCorp were on the threshold of becoming major corporations, Microsoft was still a tiny outfit developing programming languages and utilities for microcomputers. Microsoft's market was essentially limited to computer manufacturers and technically minded hobbyists. However, as described in the previous chapter, in August 1980 Gates signed a contract with IBM to develop the operating system, MS-DOS, for its new personal computer.

As sales of IBM-compatible computers grew, a copy of Microsoft's MS-DOS operating system was supplied with almost every machine. As hundreds of thousands, and eventually millions, of machines were sold, money poured into Microsoft. By the end of 1983 half a million copies of MS-DOS had been sold, netting $10 million. Microsoft's position in the software industry was unique; MS-DOS was the indispensable link between hardware and applications software that every single user had to buy.

This revenue stream enabled Microsoft to diversify into computer applications without having to rely on external venture capital. But, contrary to the mythology, Microsoft has had many more unsuccessful products than successful ones, and without the MS-DOS revenue stream it would never have grown the way it has. For example, Microsoft's first application was a spreadsheet called Multiplan, which was intended to compete with VisiCalc rather as Lotus 1-2-3 had. In December 1982 Multiplan even got a Software of the Year award, but it was eclipsed by Lotus 1-2-3. This would have been a major problem for any firm that did not have Microsoft's regular income. In mid-1982 Microsoft also began to develop a word-processing package, called Word. The product was released in November 1983 with a publicity splash that rivaled even that of Lotus 1-2-3. At a cost of $350,000 some 450,000 diskettes demonstrating the program were distributed in *PC World* magazine. Even so, Word was initially not a successful product and had a negligible impact on the market leader, WordStar. Microsoft was still no more than a medium-sized company living off the cash cow of its MS-DOS operating system.

Graphical User Interface

For the personal computer to become more widely accepted and reach a broader market, it had to become more "user-friendly." During the 1980s user-friendliness was achieved for one-tenth of computer users by using a Macintosh computer; the other nine-tenths would later achieve it through Microsoft Windows software. Underlying both systems was the concept of the *graphical user interface* (GUI, pronounced "gooey") .

Until the arrival of the Macintosh, personal computers communicated with the user through a disk operating system, or DOS. The most popular operating system for the IBM-compatible PC was Microsoft's MS-DOS. Like so much in early personal computing, DOS was derived from mainframe and minicomputer technology—in this case the notoriously efficient but intimidating Unix operating system. The early DOS-style operating systems were little better than those on the mainframes and minicomputers from which they were descended; and they were often incomprehensible to people without a computer background. Ordinary people found organizing their work in MS-DOS difficult and irritating.

The user interacted with the operating system through a "command line interface," in which each instruction to the computer had to be typed explicitly by the user, letter-perfect. For example, if one wanted to transfer a document kept in a computer file named SMITH from a computer directory called LETTERS to another directory called ARCHIVE, one had to type something like:

```
COPY A:\LETTERS\SMITH.DOC B:\ARCHIVE\SMITH.DOC
DEL A:\LETTERS\SMITH.DOC
```

If there was a single letter out of place, the user had to type the line again. The whole arcane notation was explained in a fat manual. Of course, many technical people delighted in the intricacies of MS-DOS, but for ordinary users—office workers, secretaries, and authors working at home—it was bizarre and perplexing. It was rather like having to understand a carburetor in order to be able to drive an automobile.

The graphical user interface was an attempt to liberate the user from these problems by providing a natural and intuitive way of using a computer that could be learned in minutes rather than days, for which no user manual was needed, and that would be used consistently in all applications. The GUI was also sometimes known as a WIMP interface, which stood for its critical components: windows, icons, mouse, and pull-down menus. A key idea in the new-style interface was to use the "desktop" metaphor, to which ordinary people could respond and which was far removed from technical computing. The screen showed an idealized desktop on which would be folders and documents and office tools, such as a notepad and calculator. All the objects on the desktop were represented by "icons"—for example, a document was represented by a tiny picture of a typed page, a picture of a folder represented a group of documents, and so on. To examine a document, one simply used the mouse to move a pointer on the screen to select the appropriate icon; clicking on the icon would then cause a "window" to open up on the screen so that the document could be viewed. As more documents were selected and opened, the windows would overlap, rather like the documents on a real desk would overlap one another.

The technology of user-friendliness long predated the personal computer, although it had never been fully exploited. Almost all the ideas in the modern computer interface emanated from two of the laboratories funded by the Advanced Research Projects Agency's (ARPA) Information Processing Techniques Office in the 1960s: a small human-factors research group at the Stanford Research Institute (SRI), and David Evans and Ivan Sutherland's much larger graphics research group at the University of Utah.

The Human Factors Research Center was founded at the SRI in 1963 under the leadership of Doug Engelbart, today regarded as the doyen of human-computer interaction. Engelbart had, since the mid-1950s, been struggling to get funding to develop a computer system that would act like a personal information storage and retrieval machine—in effect replacing physical documents by electronic ones and enabling the use of advanced computer techniques for filing, searching, and communicating documents. When ARPA started its computer research program in 1962 under the leadership of J. C. R. Licklider, Engelbart's project was exactly in tune with Licklider's vision of "man-computer symbiosis." ARPA funds enabled Engelbart to build up a talented group of a dozen or so computer scientists and psychologists at SRI, where they began to develop what they called the "electronic office"—a system that would integrate text and pictures in a way that was then unprecedented but is now commonly done on computers.

The modern GUI owes many of its details to the work of Engelbart's group, although by far the best-known invention is the mouse. The need for an alternative to the standard typewriter keyboard was one of the many subproblems explored in the SRI work. Several pointing devices were tried; Engelbart later recalled that "[t]he mouse consistently beat out the other devices for fast, accurate screen selection in our working context. For some months we left the other devices attached to the workstation so that a user could use the device of his choice, but when it became clear that everyone chose to use the mouse, we abandoned the other devices." All this happened in 1965; "[n]one of us would have thought that the name would have stayed with it out into the world, but the thing that none of us would have believed either was how long it would take for it to find its way out there."

In December 1968 a prototype electronic office was demonstrated by Engelbart's group at the National Computer Conference in San Francisco. Using a video projector, the computer screen was enlarged to twenty-foot width so that it could be clearly seen in the large auditorium. It was a stunning presentation, and although the system was too expensive to be practical, "it made a profound impression on many of the people" who later developed the first commercial graphical user interface at the Xerox Corporation.

The second key actors in the early GUI movement came from the University of Utah's Computer Science Laboratory where, under the leadership of David Evans and Ivan Sutherland, many fundamental innovations were made in computer graphics. The Utah environment supported a graduate student named Alan Kay, who pursued a blue-sky research project that was to result in the publication in 1969 of a highly influential computer science Ph.D. thesis. Kay's research was focused on a device, which he then called the Reactive Engine but later renamed the Dynabook, that would fulfill the personal information needs of an individual. The Dynabook was to be a personal information system, the size of a notebook, that would replace ordinary printed media. By using computer technology the Dynabook would store vast amounts of information, provide access to databases, and incorporate sophisticated information-finding tools. Of course, the system that Kay came up with for his Ph.D. program was far removed from the utopian Dynabook. Nonetheless, at the end of the 1960s, Engelbart's electronic office and Kay's Dynabook concept were the "two threads" that led to the modern graphical user interface.

What held back the practical development of these ideas in the 1960s was the lack of a compact and cost-effective technology. In the mid-1960s a fully equipped minicomputer occupied several square yards of floor space and cost up to $100,000; it was simply not reasonable to devote that much machinery to a single user. But by the early 1970s prices were falling fast, and it became feasible to exploit the ideas commercially. The first firm to do so was the Xerox Corporation, America's leading manufacturer of photocopiers.

In the late 1960s strategic planners in Xerox were becoming alarmed at the competition from Japanese photocopier manufacturers and saw the need to diversify from their exclusive reliance on the copier business. To help generate the products on which its future prosperity would depend, in 1969 Xerox established the Palo Alto Research Center (PARC) in Silicon Valley to develop the technology for the "office of the future." Approximately half of the $100 million poured into PARC in the 1970s was spent on computer science research. To head up this research, Xerox recruited Robert Taylor, a former head of the ARPA Information Processing Techniques Office. Taylor was an articulate disciple of Licklider's man-computer symbiosis vision, and his mission at PARC was to develop an "architecture of information" that would lay the foundations for the office products of the 1980s.

The concrete realization of this mission was to be a network of "Alto" desktop computers. Work on the Alto began in 1973, by which time a powerful team of researchers had been recruited—including Alan Kay from Utah, Butler Lampson and Charles Simonyi, originally from the University of California at Berkeley, and several other now well-known industry figures such as Larry Tesler. In the course of developing the Alto, Xerox PARC evolved the graphical user interface, which became "the preferred style in the 1980s just as timesharing was the preferred style of the 1970s."

The Alto computer was designed as a desktop computer with a specially constructed monitor that could display an $8\frac{1}{2} \times 11$-inch sheet of "paper." Unlike ordinary terminals, documents were not displayed as standard characters but looked like typeset documents incorporating graphical images—exactly as Kay had envisaged in the Dynabook. The Alto had a mouse, as envisaged by Engelbart, and the now-familiar desktop environment of icons, folders, and documents. In short, the system was all one would now expect from a Macintosh or a Windows-based IBM-compatible PC. However, all this was taking place in 1975, before the advent of personal computers, when "it was hard for people to believe that an entire computer" could be "required to meet the needs of one person."

Xerox decided to produce a commercial version of the Alto, the Xerox Star computer, or more prosaically the model 8010 workstation. The new computer was the most spectacular product launch of the year when it was unveiled at the National Computer Conference in Chicago in May 1981. With its eye-catching graphical user interface and its powerful office software, it was unmistakably a glimpse of the future. Yet, in commercial terms, the Xerox Star turned out to be one of the product disappointments of the decade. While Xerox had gotten everything right technically, in marketing terms it got almost everything wrong. Fundamentally, the Xerox Star was too expensive—it was hard to justify the cost of a year's salary for a powerful workstation when a simple personal computer could do much the same job, albeit less glamorously, for one-fifth of the cost. Despite being a failure in commercial terms, the Xerox Star presented a vision that would transform the way people worked on computers in the 1980s.

Steve Jobs and the Macintosh

In December 1979 Steve Jobs was invited to visit Xerox PARC. When he made his visit, the network of prototype Alto computers had just begun to demonstrate Xerox's office-of-the-future concept, and he was in awe of what he saw. Larry Tesler, who demonstrated the machines, recalled Jobs demanding, "Why isn't Xerox marketing this? . . . You could blow everybody away!" This was of course just what Xerox intended to do with the Xerox Star that was then under development.

Returning to Apple's headquarters at Cupertino, Jobs convinced his colleagues that the company's next computer would have to look like the machine he had seen at Xerox PARC. In May 1980 he recruited Larry Tesler from Xerox to lead the technical development of the new computer, which was to be known as the Lisa.

The Lisa took three years to develop. When it was launched in May 1983, it received the same kind of ecstatic acclaim that the Xerox Star had received two years earlier. The Lisa, however, was priced at $16,995 for a complete system, which took it far beyond the budget of personal-computer users and even beyond that of the corporate business machine market. Two years earlier Xerox, with its major strength in direct office sales, had not been able to make the Star successful because it was too expensive. Apple had no direct sales experience at all, and so, not surprisingly, the Lisa was a resounding commercial failure.

The failure of the Lisa left Apple dangerously exposed; its most successful machine, the aging Apple II, had been eclipsed by the IBM-compatible PC; its new machine, the Lisa, was far too expensive for the personal-computer market. What saved Apple Computer from extinction was the Macintosh.

The Macintosh project originated in mid-1979, as the idea of Jef Raskin, then manager of advanced systems at Apple. He had spent time at Xerox PARC in the early 1970s and conceived of the Macintosh as an "information appliance," which would be a simple, friendly computer that the user would simply plug in and use. Physically, he envisaged the machine as a stand-alone device with a built-in screen and a small "footprint" so that it could stand unobtrusively on the user's desk, just like a telephone. The Macintosh was named for Raskin's favorite apple—and that, as it sadly turned out for Raskin, was one of the few personal contributions to the machine he was able to make before the project was appropriated by Jobs. Raskin left Apple in the summer of 1982.

A powerful mythology has grown up around the Macintosh development. Jobs cocooned the Macintosh design group of eight young engineers in a separate building over which—only half jokingly—a pirate's flag was hoisted. Jobs had a remarkable intuitive understanding of how to motivate and lead the Macintosh design team. John Sculley, later the CEO of Apple, recalled that "Steve's 'pirates' were a hand-picked band of the most brilliant mavericks inside and outside Apple. Their mission, as one would boldly describe it, was to blow people's minds and overturn standards. United by the Zen slogan 'The journey is the reward,' the pirates ransacked the company for ideas, parts, and design plans."

The Macintosh was able to derive all the key software technology from the Lisa project. The Lisa achieved its excellent performance by using specialized hardware, which was reflected in the high price tag. Much of the design effort in the Macintosh went into obtaining Lisa-like performance at a much lower cost.

In its unique and captivating case, the Macintosh computer was to become one of the most pervasive industrial icons of the late twentieth century. As the computer was brought closer to the market, the original design team of eight was expanded to a total of forty-seven people. Jobs got each of the forty-seven to sign his or her name inside the molding from which the original Macintosh case was pressed. (Long obsolete, those original Macintoshes are now much sought after by collectors.)

In early 1983, with less than a year to go before the Macintosh launch, Steve Jobs persuaded John Sculley to become CEO of Apple. This was seen by commentators as a curious choice because the forty-year-old Sculley had achieved national prominence by masterminding the relaunch of Pepsi-Cola against Coca-Cola in the late 1970s. But behind the move lay Jobs's vision of the computer as a consumer appliance that needed consumer marketing.

In what was one of the most memorable advertising campaigns of the 1980s, Apple produced a spectacular television advertisement that was broadcast during the Super Bowl on 22 January 1984:

> Apple Computer was about to introduce its Macintosh computer to the world, and the commercial was intended to stir up anticipation for the big event. It showed a roomful of gaunt, zombie-like workers with shaved heads, dressed in pajamas like those worn by concentration camp prisoners, watching a huge viewing screen as Big Brother intoned about the great accomplishments of the computer age. The scene was stark, in dull, gray tones. Suddenly, a tanned and beautiful young woman wearing bright red track clothes sprinted into the room and hurled a sledgehammer into the screen, which exploded into blackness. Then a message appeared: "On January 24, Apple Computer will introduce the Macintosh. And you'll see why 1984 won't be like 1984."

Apple ran the commercial just once, but over the following weeks it was re-played on dozens of news and talk shows. The message was reinforced by a blaze of publicity eventually costing $15 million. There were full-page newspaper advertisements and twenty-page copy inserts in glossy magazines targeted at high-income readers.

Although priced at $2,500—only 15 percent of the cost of the Lisa—sales of the Macintosh after the first flush of enthusiasm were disappointing. Much of the hope for the Macintosh had been that it would take off as a consumer appliance, but it never did. Sculley realized that he had been misled by Jobs and that the consumer appliance idea was ill-conceived:

> People weren't about to buy $2,000 computers to play a video game, balance a checkbook, or file gourmet recipes as some suggested. The average consumer simply couldn't do something useful with a computer. Neither could the home market appreciate important differences in computer products. Computers largely looked alike and were a mystery for the average person: they were too expensive and too intimidating. Once we saturated the market for enthusiasts, it wasn't possible for the industry to continue its incredible record of growth.

Apple Computer was not the only manufacturer whose anticipation of a consumer market for personal computers was premature. In October 1983 IBM had launched a new low-cost machine, the PC Jr., hoping to attract Christmas buyers for the domestic market. The machine was so unsuccessful it had to be withdrawn the following year. If the domestic market for computers did not exist, then the only alternative was for Apple Computer to relaunch the Macintosh as a business machine.

Unfortunately the Macintosh was not well suited to the business market either. Corporate America favored an IBM-compatible personal computer for much the same reasons that an earlier generation had preferred IBM mainframes. The Macintosh fared much better in the less conservative publishing and media industries, where its powerful "desktop publishing" capabilities made it the machine of choice. The Macintosh was also popular in education, where its ease of use made it especially attractive for young children and casual student users.

To reposition the Macintosh for the business market, Apple invited several of the leading software companies to develop business applications. Not many firms had the resources to do this—and some that did, such as Lotus, found that the investment did not pay. The Macintosh never achieved much more than a 10 percent share of the computer population, and not much of that was in the profitable corporate sector. The only firm to benefit significantly from developing software for the Macintosh was Microsoft.

Microsoft had been made privy to the Macintosh project in 1981, when it was contracted to develop some minor parts of the operating software. Although Microsoft had been successful with its MS-DOS operating system for IBM-compatible computers, it had very little success in writing applications such as spreadsheets and word processors in the face of strong competitors such as Lotus and MicroPro. The Macintosh enabled Microsoft to develop a range of technologically sophisticated applications, largely insulated from the much more competitive IBM-compatible market. Later it would be able to convert the same applications so that they would run on the IBM-compatible PC. By 1987 Microsoft was deriving half of its revenues from its Macintosh software.

More important, working on the Macintosh gave Microsoft firsthand knowledge of the technology of graphical user interfaces, on which it based its new Windows operating system for the IBM-compatible PC.

Microsoft's Windows

The launch of the Macintosh in January 1984 made every other personal computer appear old-fashioned and lackluster, and it was clear that the next big thing in personal computing would be a graphical user interface for the IBM-compatible PC.

In fact, following the launch of the Xerox Star in 1981, several firms had already started to develop a GUI-based operating system for the PC, including Microsoft, Digital Research, and IBM itself. The rewards of establishing a new operating system for the PC were immense, as Microsoft had already demonstrated with MS-DOS. The firm securing the new operating system would be assured of a guaranteed stream of income that would power its future growth.

The development of a GUI-based operating system for the PC was technologically very demanding for two reasons. First, the PC was never designed with a GUI in mind, and it was hopelessly underpowered. Second, there was a strategic problem in working around the existing MS-DOS operating system; one either had to replace MS-DOS entirely with a new operating system, or else to place the new operating system on top of the old, providing a second layer of software between the user's applications and the hardware. In the first case, one would no longer be able to use the thousands of existing software applications; but in the second case, one would have an inherent inefficiency.

Perhaps the company with the strongest motive for developing a GUI-based operating system was Gary Kildall's Digital Research, the original developer of CP/M, the operating system used on 8-bit microcomputers. It was estimated that 200 million copies of CP/M were eventually sold. As noted in the previous chapter, Kildall's Digital Research failed to obtain the contract to develop the IBM PC operating system, in favor of Gates's Microsoft. Although Digital Research did develop a PC operating system called CP/M 86, it was delivered too late, it was too expensive, and by Kildall's own account it "basically died on the vine." The development of a new operating system presented a last chance for Digital Research to reestablish itself as the leading operating systems supplier.

Digital Research's GEM operating system (for Graphics Environment Manager) was launched in the spring of 1984. Unfortunately, as users quickly discovered, the system was in reality little more than a cosmetic improvement; although visually similar to the Macintosh, it lacked the capability of a full-scale operating system. Sales of GEM could not compensate for the ever-declining sales of Digital Research's CP/M operating system for the now largely obsolete 8-bit machines. In mid-1985, amid growing financial problems, Kildall resigned as CEO of Digital Research and the company he had formed faded from sight. IBM's TopView, released in 1984, fared no better. It was so slow that customers dubbed it "TopHeavy," and it became "one of the biggest flops in the history of IBM's PC business."

Microsoft Windows was the last of the new operating systems for the PC to appear. Microsoft began work on a graphical user interface project in September 1981, shortly after Gates had visited Steve Jobs at Apple and seen the prototype Macintosh computer under development. The Microsoft project was initially called the Interface Manager, but it was later renamed Windows in a neat marketing move designed "to have our name basically define the generic." It was estimated that it would take six programmer-years to develop the system. This proved to be a gross underestimate. When version 1 of Windows was released in October 1985—some two and a half years and many traumas after it was first announced—it was estimated that the program contained 110,000 instructions and had taken eighty programmer-years to complete.

Microsoft's Windows was heavily based on the Macintosh user interface—partly because it was a design on which it was very hard to improve, but also because it was felt the world would benefit from having a similar user interface in both Macintosh and Windows environments. On 22 November 1985, shortly after Windows was launched, Microsoft signed a licensing agreement with Apple to copy the visual characteristics of the Macintosh.

Although competitively priced at $99, sales of Windows were sluggish at first because it was "unbearably slow." This was true even on the most advanced PC then available, which used the new Intel 80286 microprocessor—known as the 286. Although a million copies eventually were sold, most users found the system little more than a gimmick and so the vast majority stayed with the aging MS-DOS operating system. Thus, in 1986, the PC GUI appeared to be technologically unsupportable on the existing generation of IBM-compatible computers. It would only be in the late 1980s, when the next generations of microprocessors—the Intel 386 and 486—became available that the GUI would become truly practical. Only two firms had the financial stamina to persist with a GUI-based operating system: Microsoft and IBM. In April 1987 IBM and Microsoft announced their intention to jointly develop a new operating system, OS/2, for the next generation of machines. This would be the long-term replacement for MS-DOS.

Microsoft, meanwhile, was still being carried along by the revenues from MS-DOS and was beginning to succeed in the applications market at last, with its Excel spreadsheet and Word. Gates decided that, notwithstanding OS/2, there would be short-term gains to be derived from relaunching Windows to take advantage of its own improving software technology and the faster PCs. In late 1987 Microsoft announced a new release of Windows, version 2.0. This was a major rewrite of the earlier version and had a user interface that was not merely like the Macintosh interface—as had been the case with Windows 1.0— but was for all intents and purposes identical to it. It appeared that a convergence was taking place in which, by copying the look and feel of the Macintosh, the IBM-compatible PC would become indistinguishable from it and Apple's unique marketing advantage would be lost.

Three months later, on 17 March 1988, Apple filed a lawsuit alleging that Microsoft's Windows version 2 had infringed "the Company's registered audio-visual copyrights protecting the Macintosh user interface." (Microsoft's original 1985 agreement with Apple had covered only version 1 of Windows, and Microsoft had not sought to renew the agreement for version 2.) The suit was of enormous importance for the future of the personal-computer industry. If it were upheld, then it would have a severe negative effect on the great majority of users who did not use Macintosh computers and would be interpreted by developers that all user interfaces had to be different. If Apple were to succeed in the suit, it would be like a car manufacturer being able to copyright the layout of its instruments and every manufacturer having to ensure that its autos had a novel and different instrument layout. This would clearly be against the public interest. Although the target of Apple's suit was Microsoft, it was argued by some that the biggest loser would be IBM, with its OS/2 operating system. If IBM was prevented from marketing a Macintosh-like computer, then the entire PC industry would be trapped in a cul-de-sac.

The lawsuit was to rattle on for three more years before it was finally dismissed. Meantime, while the law ran its leisurely pace, Microsoft was achieving one of the most dramatic growth spurts of any business in the twentieth century, and Gates was making history by becoming the world's youngest self-made billionaire. Some 2 million copies of Windows had been sold by early 1989, massively outselling IBM's OS/2 operating system, which had been launched in early 1988 (and in which Microsoft no longer expressed any interest). It was moot which of the systems was technically superior; the important factor was that Microsoft's marketing was much better.

Much of the software industry's rapid growth in the late 1980s, including that of Microsoft, was due to the ever-increasing sophistication of its products. Applications packages were routinely updated every eighteen months or so, generating additional revenues as users traded up for more sophisticated packages, making better use of rapidly improving hardware. Windows was no exception. Undeterred by the Apple-Microsoft lawsuit, by mid-1990 Windows was ready for yet another relaunch. It now consisted of 400,000 lines of code, nearly four times the size of the first version released in 1985. In total it was estimated to have cost $100 million to develop.

On 22 May 1990 Windows 3.0 was launched around the world:

> [S]ome 6,000 people were on hand as the City Center Theater in New York City became center stage for the multimedia extravaganza. Gala events took place in 7 other North American cities, which were linked via satellite to the New York stage for a live telecast, and in 12 major cities throughout the world, including London, Amsterdam, Stockholm, Paris, Madrid, Milan, Sydney, Singapore, and Mexico City. The production included videos, slides, laser lights, "surround sound," and a speech by Bill Gates, who proclaimed Windows 3 "a major milestone" in the history of software, saying that it "puts the 'personal' back into millions of MS-DOS-based computers."

Ten million dollars was spent on a blaze of publicity for Windows. By Microsoft's Bill Gates's own account, it was "the most extravagant, extensive, and expensive software introduction ever."

In a way that only became fully apparent in the mid-1990s following the launch of Windows 95, Microsoft had come to own the "platform" of personal computing. This would give Microsoft an industry prominence comparable with that of IBM in its heyday.

CD-ROMs and Encyclopedias

As well as software, users of personal computers wanted information, ranging from reference works such as encyclopedias and dictionaries to multimedia entertainment. By the late 1990s such information would be routinely obtained from the Internet, but even though consumer networks existed in the mid-1980s (discussed later in this chapter), the speed and capacity of these networks was simply too limited for supplying large volumes of information. The principal bottleneck was the modem, the device that connected a home computer to a network through a telephone line. To transmit, for example, an article of about the length of a chapter in this book would have taken thirty minutes; add a few images and that time would double. Until inexpensive high-speed communications arrived, the CD-ROM played an historic role in the way large volumes of information were shipped to home computers.

The CD-ROM (compact disc—read-only memory) was the outcome of a joint research development by Sony and Philips in the early 1980s, and devices were first marketed in 1984. The CD-ROM was based on the technology of the audio CD and offered a storage capacity of over 500 megabytes—several hundred times that of a floppy disk. Although CD-ROM drives were expensive on their introduction, costing $1,000 or more for several years, their huge capacity gave them the potential to create an entirely new market for computer-enabled content that the embryonic computer networks would not be able to satisfy for another fifteen years.

As long as the price of a CD-ROM drive remained around $1,000, its use was limited to corporations and libraries, mainly for business information and high-value publications. In the consumer arena, two of the industry leaders in personal-computer software, Gary Kildall and Bill Gates, played critical roles in establishing a market for CD-ROM media. Both Kildall and Gates hit on the idea of the CD-ROM encyclopedia as the means to establish the home market for CD-ROMs. Kildall explained: "Everyone knows encyclopedias usually cost about $1,000. Someone can rationalize buying a computer that has the encyclopedia, if it's in the same price range as the printed encyclopedia."

While still running Digital Research, Kildall started a new CD-ROM publishing operation as a separate company. He secured the rights to the *Grolier*

Encyclopedia, a venerable junior encyclopedia established shortly after World War I. Grolier's *Academic American Encyclopedia* on CD-ROM was released in 1985, well ahead of any similar product. Priced at several hundred dollars, the Grolier CD-ROM can best be described as worthy but dull. It sold well to schools but did not induce consumers to invest in CD-ROM equipment.

Microsoft took much longer to get an encyclopedia onto the market. As early as 1985, Gates had tried to get the rights to the *Encyclopedia Britannica*, the leader in both content and brand. However, Encyclopedia Britannica Inc. was wary of cannibalizing its lucrative hard-copy sales, and in any case it already had a CD-ROM version of its lesser-known *Compton's Encyclopedia* under way. Next, Microsoft tried the *World Book Encyclopedia*, but it too had its own CD-ROM project in the pipeline.

Microsoft gradually worked its way down the pantheon of American encyclopedias, finally obtaining the rights to *Funk and Wagnalls New Encyclopedia* in 1989. *Funk and Wagnalls* had an unglamorous but profitable niche as a $3 per volume impulse purchase in supermarkets. Microsoft's decision to acquire the rights to *Funk and Wagnalls* has frequently been derided as pandering to the educationally challenged working class, a criticism more than a little tinged by academic snobbery. But it was in truth very much a junior encyclopedia of negligible authority. Microsoft, however, proceeded to sculpt this unpromising material into *Encarta*, adding video and sound clips and other multimedia elements to the rather thin content. Released in early 1993 at a price of $395, it sold only modestly.

In fact, the CD-ROM encyclopedia did not turn out to be the killer application that would start the CD-ROM revolution. Quite independently, the cost of CD-ROM drives had fallen slowly but continuously over a period of several years, to a price of about $200 by 1992—comparable to that of a printer or a hard disk drive. With the fall in price of CD-ROM drives, there was at last the possibility of a mass market for CD-ROM media. In 1992 the recreational software publisher Broderbund released a title *Grandma and Me* for $20 that became an international bestseller and probably the best-known CD-ROM title of its time. Thousands of other titles from hundreds of publishers soon followed. At about the same time publishers of video games started to release CD-ROM titles with multimedia effects. The price of CD-ROM encyclopedias, so recently priced at several hundred dollars, fell into line. The price of *Encarta* was cut to $99 for the 1993 Christmas season, and its competitors followed suit.

The availability of low-cost CD-ROM encyclopedias had a devastating effect on the market for traditional hard-bound encyclopedias. As a *Forbes* magazine put it:

How long does it take a new computer technology to wreck a 200 year-old publishing company with sales of $650 million and a brand name recognized all over the world? Not very long at all. There is no clearer, or sadder, example of this than how the august Encyclopedia Britannica Inc. has been brought low by CD-ROM technology.

In 1996, under new ownership, a CD-ROM edition of the *Encyclopedia Britannica* finally hit the market for less than $100.

Consumer Networks

It sometimes seems like the Internet arrived from nowhere in the early 1990s, but in the 1980s there were vibrant consumer networks designed for computer enthusiasts.

The leading consumer network throughout the 1980s was CompuServe. The network had originally been set up in 1969, as an old-line time-sharing service of the type described in chapter 9 by Jeff Wilkins, a twenty-five-year-old entrepreneur trained in electrical engineering. It was the subsidiary of an Ohio-based insurance company that happened to be run by Wilkins's father-in-law. Unfortunately, CompuServe was established just as the fad for computer utilities was on the wane. Only after two years of heavy losses did it begin to turn a modest profit, selling time-sharing services to the insurance industry.

Rather like electric utilities, business-oriented time-sharing services suffered from wildly fluctuating demand. During nine-to-five office hours, the service was used to capacity, but in the evening when the business day was over, and on the weekends, the system was heavily underused. In 1978, with the arrival of personal computers, Wilkins saw an opportunity to sell this spare capacity to computer hobbyists. Working with the Midwest Association of Computer Clubs—which distributed the access software to its members—he created a service called MicroNet. Subscribers could access the system at off-peak times for a few dollars an hour. MicroNet proved very popular and Wilkins decided to extend the service nationally. However, he first had to get the access software into the hands of consumers nationwide. To do this he made an arrangement with the Tandy Corporation to distribute a $39.95 "starter kit" in its 8,000 Radio Shack stores. The kit included a user manual, access software, and $25 worth of free time. (The deal worked well for Tandy, too, as it had just introduced its TRS 80 Color Computer, and the availability of CompuServe gave it a unique selling point.)

Unlike CompuServe's corporate customers, who mainly wanted financial analysis programs and insurance services, home-computer users were much more interested in "content" and in communicating with other subscribers. Newspapers were signed up—starting with the *Columbus (Ohio) Dispatch* in 1980, but eventually including major newspapers such as the *New York Times*. Computer games were provided, often written by subscribers who were recompensed on a royalty basis. Electronic mail was provided so that subscribers could send messages to one another. A particularly popular feature (later known as a chat room) was the CB Channel, named in homage to the Citizens Band radio fad of the early 1980s. This allowed several users to communicate in a "forum" in real time. By summer 1984, CompuServe had 130,000 subscribers in 300 cities, served by twenty-six mainframe computers and a staff of 600 in its Columbus headquarters:

> A CompuServe subscriber with a personal computer and a telephone modem can call in and use the service to make an airline reservation, read the *Washington Post*, scan the AP Wire Service, pick up some free software, check a stock portfolio, or browse a data bank or encyclopedia. There are also a number of popular "bulletin boards," which allow people to post messages for one another. One functions like a CB radio, allowing up to 40 users at a time to chew the fat. The basic fee is $12.50 an hour ($6 nights and weekends). Many of the special services cost extra, but an experienced subscriber can accomplish a lot while drinking a cup of coffee. It's also still possible to use CompuServe for its original purpose, time-sharing on a large computer.

CompuServe was the most successful early consumer network, and for several years the number of subscribers exceeded those of all the other services put together. As PCs moved into the workplace, CompuServe also tapped the business market, offering Executive Services—primarily access to conventional databases such as Standard & Poors' company information service and Lockheed's Dialog information database. Users paid a premium for these services, which was shared between CompuServe and the information provider.

Although only about one in twenty computer owners used an on-line service in the mid-1980s, they were widely perceived as an opportunity with tremendous potential, and CompuServe began to attract some major competitors. The two most important were The Source and Prodigy. The Source was a network set up by the Readers Digest Association and the computer firm CDC, while Prodigy was created by Sears and IBM. In each case, one partner saw the network as an extension to its existing business while the other provided the computing know-how. Thus Readers Digest Association saw consumer networks as a new form of publishing (which in time it became, though not until long after the association had quit the business); Sears saw the network as an extension of its famous mail-order catalog. Both organizations also had excellent distribution channels through which they could get access software into the hands of subscribers. But success was elusive, and it proved very difficult to compete with CompuServe. The Source was sold to CompuServe in 1989, while Prodigy incurred huge losses before its owners abandoned it in 1996.

CompuServe's relative success is perhaps explained by what economists call network effects. Rather like the telephone system, it is only the existence of large numbers of subscribers that makes the service useful. If only one individual had a telephone it would be no use at all, as there would be no one to talk to. But when there were a thousand subscribers the system became useful; when there were millions of users it became indispensable. In the same way, the bigger the computer network, the more useful it was to subscribers. At a time when users of one network could not communicate with those of another, it made sense to belong to the biggest network. By 1990, CompuServe had 600,000 subscribers and it offered literally thousands of services: from home banking to hotel reservations, from Roger Ebert's movie reviews to *Golf* magazine. In 1987 CompuServe had extended into Japan (with a service called NiftyServe), and in the early 1990s it began a service in Europe.

CompuServe did, however, prove vulnerable to one newcomer, America On-Line (AOL). AOL had begun as a me-too consumer network in the mid-1980s. Its success was fairly modest until it developed a service for the Macintosh computer. Rather in the way the Macintosh carved a successful niche with its graphical user interface, the AOL service for the Macintosh benefited from developing a friendly interface that set it apart from the competition. In 1991 the service was extended to the IBM-compatible PC, providing "an easy-to-use point-and-click interface anyone's grandmother could install." As always with consumer networks, a major barrier was getting access software into the hands of consumers and inducing them to try the service. AOL hit on the idea of inserting a floppy disc with access software in computer magazines, together with a free trial period. AOL became (and remains) famous for "carpet bombing" America with floppy discs, and later CD-ROMs. By making the service free to try out, hundreds of thousands of consumers did just that. And because the service was user-friendly, many of them stayed on as paying customers. True, an even greater number did not stay, creating a phenomenon known as "churn"—where as many subscribers were said to be leaving as joining. Nonetheless by the end of 1995 AOL claimed it had 4.5 million subscribers (including many in Europe where, surprisingly, AOL did not feel the need to make its name less imperialistic).

Although the Internet was eventually to become the universal on-line medium, in the early 1990s this was not foreseeable. At that time all the leading consumer services were large proprietary networks that had sufficient size and content to be self-contained worlds. The main reason a subscriber might want to use a different network would be to exchange e-mail with an individual who happened to belong to a competing network. This was a relatively simple technological problem that was being addressed.

By the early 1990s networking was in the air. There was much political talk of "information superhighways" and many—not yet a majority by any means—but several million home-computer owners were using one of the consumer networks. Microsoft, always the most opportunistic of the personal-computer software firms, decided to enter the market for on-line services about 1992. One option was to buy up one of the "big three" networks—AOL, CompuServe, or Prodigy. But none of these was a possible acquisition, either because the business was not for sale, or because it did not make a good cultural fit. Microsoft therefore decided to build its own network from scratch, to be called the Microsoft Network, or MSN. It would be a proprietary network with the usual range of on-line services: chat rooms, e-mail, electronic versions of popular publications, stock prices, software downloads, and so on. Work started in mid-1993 and rapidly gathered momentum. Hundreds of information providers became enthusiastic recruits—if Microsoft succeeded with MSN the way it seemed to succeed with everything else, no one wanted to be left out.

As noted earlier, a major barrier to starting up an on-line service was getting the access software into the hands of consumers. CompuServe had used the Radio Shack stores, while AOL has used magazine inserts. For Microsoft, the upcoming launch of its new Windows 95 operating system presented a perfect opportunity to bundle the access software with the operating system itself. A mouse click on the MSN icon would then transport the user to the network. Inevitably, and not unreasonably, there were howls of protest from the other consumer networks who demanded a level playing field—that their access software should be equally prominent on the Windows desktop. In fact, they need not have worried. The Microsoft network was a loss-maker from the start. It had taken about three years to set up MSN under the old proprietary network model, and the world had changed around it. As a historian of Microsoft explained:

> In 1993, the media had carried stories about the information highway, ad nauseam. By 1994, journalists had shifted their attention to the Internet. By 1995, they wrote about nothing else. From academic obscurity to media darling, the

Internet's arrival on the front covers of news magazines meant that not just MSN but the other commercial on-line services would be directly affected.

Indeed. The Internet was to become overwhelmingly the most important development in computing from the mid-1990s onward. But this was only possible because in the previous decade the personal computer had been transformed from an unfriendly technological system into something approaching an information appliance.

In that decade, software companies provided application programs that transformed the general-purpose personal computer into a special-purpose machine that met the needs of particular users—whether they were business analysts using spreadsheets, individuals word processing documents, or players of games. While many of these applications simplified using a computer—for example, by making a word-processor program appear like a familiar typewriter—it was the graphical user interface that opened up the computer to the wider society. Instead of typing complicated commands, users could now simply point and click their way through the most complex software, while knowing essentially nothing about the inner workings of a computer.

Although the computer of the mid-1980s had become easier to use, it was an isolated machine. It was the modern replacement for the typewriter or electronic calculator and even a popular new item in the games room—but it was no substitute for a library or a telephone. To become truly useful, the personal computer needed access to large quantities of information comparable with that of a modest library, and it needed the ability to communicate with other computers. CD-ROM technology and consumer networks played complementary roles in this transformation. CD-ROM technology provided information in high volume, but it was information that was static and rapidly became obsolete. Consumer networks held large volumes of information, but users could only obtain it in tiny quantities as it trickled through the bottleneck of a modem. In the early years of the twenty-first century, the rise of the Internet and high-speed broadband connections would at last provide users with the information they needed at an acceptable speed.

12

:: FROM THE WORLD BRAIN
TO THE WORLD WIDE WEB

IN THE early 1990s the Internet—the system that links millions of computers around the world—became big news. In the fall of 1990, there were just 313,000 computers attached to the Internet; five years later, the number was approaching 10 million; and by the end of 2000 the number exceed 100 million. Although computer technology is at the heart of the Internet, its importance is economic and social; it gives computer users the ability to communicate, to gain access to information sources, and to conduct business.

The Internet sprang from a confluence of three desires, two that emerged in the 1960s and one that lay much further back in time. First, there was the rather utilitarian desire for an efficient, fault-tolerant networking technology suitable for military communications that would never break down. Second, there was a wish to unite the world's computer networks into a single system. Just as the telephone would never have become the dominant person-to-person communications medium if users had been restricted to the network of their particular provider, so the world's isolated computer networks would be far more useful if they were joined together. But the most romantic ideal—perhaps dating as far back as the Library of Alexandria in the ancient world—was to make readily available the world's store of knowledge.

From the Encyclopedia to the Memex

The idea of making the world's store of knowledge available to the ordinary person is a very old dream. It was the idea that drove an Englishman to create a two-volume *Cyclopaedia* (1728) and a French publisher to ask the philosopher Denis

Diderot to translate it. In the end, Diderot went much further, and created the first great encyclopedia in the eighteenth century. The multivolume *Encyclopédie* was one of the central projects of the Age of Enlightenment, which tried to bring radical and revolutionary reform by giving knowledge and therefore power to the people. The *Encyclopédie* was in part a political act; similarly, the Internet has a political dimension. The *Encyclopedia Britannica,* modeled directly on the *Encyclopédie,* appeared in 1768. Of course, neither the *Encyclopédie* nor the *Encyclopedia Britannica* could contain *all* of the world's knowledge. But they contained a significant portion of it, and—at least as important—they brought order to the whole universe of knowledge, giving people a sense of what there was to know.

The nineteenth century saw an explosion in the production of human knowledge. In the early decades of the century, it had been possible for a person of learning to be comfortable with the whole spectrum of human knowledge, in both the arts and sciences. For example, Mark Roget—now famous only as the compiler of the thesaurus—earned his living as a physician, but he was also an amateur scientist and a member of the Royal Society of London, an educationalist, and a founder of the University of London. Charles Babbage, besides being famous for his calculating engines, was an important economist; he also wrote on mathematics and statistics, geology and natural history, and even theology and politics.

By the twentieth century, however, the enormous increase in the world's knowledge had brought about an age of specialization in which it became very unusual for a person to be equally versed in the arts and the sciences and impossible for anyone to have a deep knowledge of more than a very narrow field of learning. It was said, for example, that by 1900 no mathematician could be familiar even with all the different subdisciplines of mathematics.

In the years between the two world wars a number of leading thinkers began to wonder whether it might be possible to arrest this trend toward specialization by organizing the world's knowledge systematically so that, at the very least, people could once again know what there was to know. The most prominent member of this movement was the British socialist, novelist, and science writer Herbert George Wells, best known in the United States as the author of *The War of the Worlds* and *The Time Machine.* During his own lifetime, Wells had seen the world's store of knowledge double and double again. He was convinced that narrow specialization, and the failure of even educated people to understand more than a tiny fraction of the world's knowledge, was causing the world to descend into barbarism in which people of learning were being "pushed aside by men like Hitler." During the 1930s he wrote pamphlets and articles and gave speeches about his project for a World Encyclopedia that would do for the twentieth century what Diderot had done for the eighteenth. Wells failed to interest publishers because of the enormous cost of such a project, and so in the fall of 1937 he embarked on a U.S. lecture tour hoping to raise funds.

He covered five cities, and his last presentation, in New York, was also broadcast on the radio. In his talk, titled "The Brain Organization of the Modern World," Wells explained that his World Encyclopedia would not be an encyclopedia in the ordinary sense:

> A World Encyclopedia no longer presents itself to a modern imagination as a row of volumes printed and published once for all, but as a sort of mental clearing house for the mind, a depot where knowledge and ideas are received, sorted, summarized, digested, clarified and compared. . . . This Encyclopedic organization need not be concentrated now in one place; it might have the form of a network [that] would constitute the material beginning of a real World Brain.

Wells never explained how his "network" for the World Brain would be achieved, beyond supposing that it would be possible to physically store all the data on microfilm. All the information in the world would be no good, however, if it were not properly organized. He thus envisaged that a "great number of workers would be engaged perpetually in perfecting this index of human knowledge and keeping it up to date."

During his tour, Wells was invited to lunch as the distinguished guest of President Franklin Roosevelt. Wells lost no time in trying to interest his host in the World Brain project, but, perhaps not surprisingly, Roosevelt had more pressing problems. Wells left the lunch a disappointed man. Time was running out for the World Brain; as the world drifted into World War II, he was forced to abandon the project and fell into a depression that never lifted. He lived to see the end of the war but died the following year at the age of eighty.

Wells's dream did not die with him. In the postwar period the idea resurfaced and was given new vigor by Vannevar Bush, the scientist and inventor who had developed analog computers and risen to become chief scientific adviser to the president and head of the Office of Scientific Research and Development, where he directed much of the U.S. scientific war effort.

In fact, Bush had first proposed an information storage machine several years before the war. This was to be a desk-like device that could hold the contents of a university library "in a couple of cubic feet." With the start of the war

Bush had to put these ideas to one side, but in its final months he turned his mind to what scientists might do in the postwar era. For him, one problem stood out above all others: coping with the information explosion. In July 1945 he set his ideas down in a popular article, "As We May Think," for *Atlantic Monthly*. A few weeks later it reached a wider readership by being reprinted in *Life* magazine. This article established a dream that has been pursued by information scientists for more than fifty years.

During the war, as Wells had predicted, microfilm technology had advanced to a point where the problem of storing information was essentially solved. Using microfilm technology, Bush noted, it would be possible to contain the *Encyclopedia Britannica* in a space the size of a matchbox, and "a library of a million volumes could be compressed into one end of a desk." Thus the problem was not so much *containing* the information explosion as being able to make use of it. Bush envisaged a personal information machine that he called the memex:

> A memex is a device in which an individual stores all his books, records, and communications, and which is mechanized so that it may be consulted with exceeding speed and flexibility. It is an enlarged intimate supplement to his memory.
>
> It consists of a desk, and while it can presumably be operated from a distance, it is primarily the piece of furniture at which he works. On the top are slanting translucent screens, on which material can be projected for convenient reading. There is a keyboard, and sets of buttons and levers. Otherwise it looks like an ordinary desk.

The memex would allow the user to browse through information:

> If the user wishes to consult a certain book, he taps its code on the keyboard, and the title page of the book promptly appears before him, projected onto one of his viewing positions. Frequently-used codes are mnemonic, so that he seldom consults his code book; but when he does, a single tap of a key projects it for his use. Moreover, he has supplemental levers. On deflecting one of these levers to the right he runs through the book before him, each page in turn being projected at a speed which just allows a recognizing glance at each. If he deflects it further to the right, he steps through the book 10 pages at a time; still further at 100 pages at a time. Deflection to the left gives him the same control backwards.
>
> A special button transfers him immediately to the first page of the index. Any given book of his library can thus be called up and consulted with far

greater facility than if it were taken from a shelf. As he has several projection positions, he can leave one item in position while he calls up another. He can add marginal notes and comments . . . just as though he had the physical page before him.

Bush did not quite know how, but he was sure that the new technology of computers would be instrumental in realizing the memex. He was one of the very few people in 1945 who realized that computers would one day be used for something more than rapid arithmetic.

In the twenty years that followed Bush's article, there were a number of attempts at building a memex using microfilm readers and simple electronic controls, but the technology was too crude and expensive to make much headway. When Bush revisited the memex idea in 1967 in his book *Science Is Not Enough*, his dream of a personal information machine was still "in the future, but not so far."

As he wrote those words, the intellectual problems of constructing a memex-type information system using computer technology had, in principle, been largely solved. J. C. R. Licklider, the head of the Advanced Research Project Agency's (ARPA) Information Processing Techniques Office, for one, was working as early as 1962 on a project he called the Libraries of the Future, and he dedicated the book he published with that title: "however unworthy it may be, to Dr. Bush." In the mid-1960s Ted Nelson coined the term *hypertext*, and Douglas Engelbart was working on the practical realization of similar ideas at the Stanford Research Institute. Both Nelson and Engelbart claim to have been directly influenced by Bush. Engelbart later recalled that, as a lowly electronics technician in the Philippines during World War II, he "found this article in *Life* magazine about his memex, and it just thrilled the hell out of me that people were thinking about something like that. . . . I wish I could have met him, but by the time I caught onto the work, he was already in a nursing home and wasn't available."

While the intellectual problems of designing a memex-type machine were quickly solved, establishing the physical technology—a network of computers—would take a long time.

The ARPANET

The first concrete proposal for establishing a geographically distributed network of computers was made by Licklider in his 1960 *Man-Computer Symbiosis* paper:

It seems reasonable to envision, for a time 10 or 15 years hence, a "thinking center" that will incorporate the functions of present-day libraries together with anticipated advances in information storage and retrieval. . . . The picture readily enlarges itself into a network of such centers, connected to one another by wide band communication lines and to individual users by leased-wire services. In such a system, the speed of the computers would be balanced, and the cost of the gigantic memories and the sophisticated programs would be divided by the number of users.

In 1963, when the first ARPA-sponsored time-shared computer systems became operational, Licklider set into action a program he privately called his Intergalactic Computer Network, publicly known as the ARPANET.

The stated motive for ARPANET was an economic one. By networking ARPA's computer systems together, the users of each computer would be able to use the facilities of any other computer on the network; specialized facilities would thus be available to all, and it would be possible to spread the computing load over many geographically separated sites. For example, because the working day for East Coast computer users started several hours ahead of the West Coast's, they could make use of idle West Coast facilities in the morning; and when it was evening on the East Coast, their positions would be reversed. ARPANET would behave rather like a power grid in which lots of power plants worked in harmony to balance the load.

In July 1964, when Licklider finished his two-year tenure as head of ARPA's Information Processing Techniques Office (IPTO), he had a powerful say in appointing his successors, who shared his vision and carried it forward. His immediate successor was Ivan Sutherland, an MIT-trained graphics expert, then at the University of Utah, who held office for two years. He was followed by Robert Taylor, another MIT alumnus, who later became head of the Xerox PARC computer science program.

Between 1963 and 1966 ARPA funded a number of small-scale research projects to explore the emerging technology of computer networking. ARPA was not the only organization sponsoring such activity; there were several other groups interested in computer networking in the United States, England (especially at the National Physical Laboratory), and France. By 1966 the computer networking idea was ready for practical exploitation, and IPTO's head, Robert Taylor, decided to develop a simple experimental network. He invited a rising star in the networking community, Larry Roberts, to head up the project.

Larry Roberts had earned his Ph.D. at MIT in 1963. He was first turned on to the subject of networking by a discussion with Licklider at a conference in November 1964, and "his enthusiasm infected me." When he was approached by Taylor to head IPTO's networking program, Roberts was working on an IPTO-sponsored networking project at MIT's Lincoln Laboratory. It was a wonderful opportunity, but Roberts was initially reluctant to exchange an active research role at the Lincoln Laboratory for what would likely prove a bureaucratic one with IPTO. However, Taylor went over Roberts's head and reminded his boss, the head of the Lincoln Laboratory, that the laboratory secured more than 50 percent of its funding from ARPA. Roberts, who was quickly learning how the world of government funding operated, knew where his duty and best career move lay.

When he took over the ARPANET project in 1966, Roberts had to find solutions to three major technical problems before work could begin on a practical network. The first problem was how to physically connect all the time-sharing systems together. The difficulty here was that, if every computer system had to be connected to every other, the number of communications lines would grow geometrically. Networking the seventeen ARPA computers that were then in existence would require a total of 136 ($17 \times 16 \div 2$) communications lines. The second problem was how to make economic use of the expensive high-speed communications lines connecting computers. Experience with commercial time-sharing computers had already shown that less than 2 percent of the communications capacity of a telephone line was productively used because most of a user's time was spent thinking, during which the line was idle. This did not matter too much with local phone lines, but it would be insupportable on high-speed long-distance lines. The third problem Roberts faced was how to link together all the computer systems, which came from different manufacturers and had many varieties of operating software that had taken several years to develop. Enough was known about the software crisis at this stage to want to avoid the extensive rewriting of operating systems.

Unknown to Roberts, a solution to the first two problems had already been invented. Known as "store-and-forward packet switching," the idea was first put forward by Paul Baran of the RAND Corporation in 1961 and was independently reinvented in 1965 at the National Physical Laboratory in England by Donald Davies, who coined the term *packet switching*. Davies recognized the packet-switching concept to be similar to an older telegraph technology.

In telegraph networks, engineers had already solved the problem of how to avoid having every city connected to every other. Connectivity was achieved by using a number of switching centers located in major cities. Thus to send a telegram from, say, New York to San Francisco, the message might pass through intermediate switching centers in Chicago and Los Angeles before arriving in

San Francisco. In the early days of the telegraph, during the late nineteenth century, at each switching center an incoming telegram would be received on a Morse sounder and written out by a telegraph clerk. It would then be retransmitted by a second telegraph clerk to the next switching center. This process would be repeated at each switching center, as the telegram was relayed across the country to its final destination.

An incidental advantage of creating a written copy of the telegram was that it could act as a storage system, so that if there was a buildup of traffic, or if the onward switching center was too busy to receive the message, it could be held back until the lines became quieter. This was known as the store-and-forward principle. In the 1930s these manual switching centers were mechanized in "torn-tape offices," where incoming messages were automatically recorded on perforated paper tape and then retransmitted mechanically. In the 1960s the same functions were being computerized using disk stores instead of paper tape as the storage medium.

Store-and-forward packet switching was a simple elaboration of these old telegraph ideas. Instead of having every computer connected to every other, store-and-forward technology would be used to route messages through the network; there would be a single "backbone" communications line that connected the computers together, with other connections being added as the need arose. Packet-switching technology addressed the problem of making economic use of the high-speed communications lines. So that a single user did not monopolize a line, data would be shuttled around the network in packets. A packet was rather like a short telegram, with each packet having the address of the destination. A long message would be broken up into a stream of packets, which would be sent as individual items into the network. This would enable a single communications line to carry many simultaneous human-computer interactions by transmitting packets in quick succession. The computers that acted as the switching centers—called *nodes* in the ARPANET—would simply receive packets and pass them along to the next node on the route toward the destination. The computer at the destination would be responsible for reconstituting the original message from the packets. In effect, by enabling many users to share a communications line simultaneously, packet switching did for telecommunications what time-sharing had done for computing.

All of this was unknown to Roberts until he attended an international meeting of computer network researchers in Gatlinburg, Tennessee, in October 1967. There he learned of the packet-switching concept from one of Donald Davies's English colleagues. He later described this as a kind of revelation: "Suddenly I learned how to route packets."

The final problem that remained for Roberts was how to avoid the horrendous software problems of getting the different computers to handle the network traffic. Fortunately, just as Roberts was confronting this problem, the first minicomputers had started to come onto the market, and the solution came to him in a eureka moment during a taxicab ride: the Interface Message Processor (IMP). Instead of the existing software having to be modified at each computer center, a separate inexpensive minicomputer, the IMP, would be provided at every node to handle all the data-communications traffic. The software on each computer system—called a host—would need only a relatively simple modification to collect and deliver information between itself and the IMP. Thus there was only one piece of software to worry about—a single software system to be used in all the IMPs in the network.

The ARPANET project, from having been little more than a gleam in IPTO's eye when Roberts took it over in 1966, had become concrete. In the summer of 1968 he succeeded Taylor as director of IPTO, and the ARPANET project went full steam ahead. Work began on a $2.5 million pilot scheme to network together four computer centers—at the University of California at Los Angeles, the University of California at Santa Barbara, the University of Utah, and the Stanford Research Institute. The software for the IMPs was contracted out to Bolt, Beranek, and Newman (BBN), while on the university campuses a motley collection of graduate students and computer center programmers worked on connecting up their host computers to the IMPs. The participation of so many graduate students in developing ARPANET, typically working on a part-time basis as they completed their computer science graduate studies, created a distinctive and somewhat anarchic culture in the network community. This culture was as strong in its way as the computer-amateur culture of the personal-computer industry in the 1970s. But unlike the personal-computer world, this culture was much more persistent, and it accounted for the unstructured and anarchic state of the early Internet.

By 1970 the four-node network was fully operational and working reliably. The other ARPA-funded computer centers were soon joined up to the network, so that by the spring of 1971 there were twenty-three hosts networked together. Although impressive as a technical accomplishment, ARPANET was of little significance to the tens of thousands of ordinary mainframe computers in the world beyond. If the networking idea was ever to move beyond the ARPA community, Roberts realized that he had to become not merely a project manager but an evangelist. He proselytized ARPANET to the technical community at academic conferences but found he was preaching to the converted. He therefore decided to organize a public demonstration of the ARPANET that simply no one would be able to ignore, at the first International Conference on Computer Communications (ICCC) in Washington in the fall of 1972.

The conference was attended by over a thousand delegates, and from a demonstration area equipped with forty terminals people were able to directly use any one of more than a dozen computers, ranging from Project MAC at MIT to the University of California computer center. A hookup was even arranged with a computer in Paris. Users were able to undertake serious computing tasks such as accessing databases, using meteorological models, and exploring interactive graphics; or they could simply amuse themselves with an air-traffic simulator or a chess game. The demonstration lasted three days and left an indelible impression on those who attended. One participant noted: "There was more than one person exclaiming, 'Wow! What is this thing?'"

The ICCC demonstration was a turning point for both the ARPANET and for networking generally. The technology had suddenly become real, and many more research organizations and universities clamored to become connected to ARPANET. A year after the conference, there were forty-five hosts on the network; four years later there were 111. It was an improvement, but growth was still slow.

The Popularity of E-mail

It was not, however, the economics of resource sharing, the ability to use remote computers, or even the pleasure of playing computer games that caused the explosion of interest in networking; indeed, most users never made use of any of these facilities. Instead it was the opportunity for communicating through electronic mail that attracted users.

Electronic mail had never been an important motivation for ARPANET. Indeed, no electronic mail facilities were provided initially, even though the idea was by no means unknown. An e-mail system had been provided on MIT's Project MAC in the mid-1960s, for example; but because users could only mail other MIT colleagues, it was never much more than an alternative to the ordinary campus mail.

In July 1971 two BBN programmers developed an experimental mail system for ARPANET. According to one of the BBN network team:

> When the mail [program] was being developed, nobody thought at the beginning it was going to be the smash hit that it was. People liked it, they

thought it was nice, but nobody imagined that it was going to be the explosion of excitement and interest that it became.

Electronic mail soon exceeded all other forms of network traffic on ARPANET, and by 1975 there were over a thousand registered e-mail users. The demand for e-mail facilities was also a major driving force for the first non-ARPA networks. One of the most important of these was Usenet, a network formed in 1978 for colleges that had been excluded from connection to ARPANET. An unplanned spin-off from this network was the Usenet news system. Acting like a giant electronic bulletin board, the system enabled network users to subscribe to news groups where like-minded individuals could exchange views. Initially designed by a student at Duke University and another at the University of North Carolina to exchange news, the idea quickly caught on. At first news groups exchanged information mainly about computer-related matters, but eventually they included thousands of different topics. By 1991 there were 35,000 nodes on the Usenet system and millions of subscribers to the news network.

However, while most people quickly tired of using news groups, networked computing and e-mail became an integral part of the modern way of doing business, and the existing computer services industry was forced to respond. First, the time-sharing firms began to restructure themselves as network providers. BBN's Telcomp time-sharing service, for example, relaunched itself as the Telnet network in 1975 (with Larry Roberts as CEO). Telnet initially established nodes in seven cities; and by 1978 there were nodes in 176 U.S. cities and fourteen foreign countries. In 1979 Telcomp's rival Tymshare relaunched itself as Tymnet. Established communications firms such as Western Union and MCI were also providing e-mail services, and the business press was beginning to talk of the Great Electronic Mail Shootout. In the public sector, government organizations such as NSF and NASA also developed networks, while in the education sector consortia of colleges developed networks such as Merit, Edunet, and Bitnet, all of which became operational in the first half of the 1980s.

Electronic mail was the driving force behind all these networks. It became popular because it had so many advantages over conventional long-distance communications; and these advantages were consolidated as more and more networks were established—the more people there were on the networks, the more useful e-mail became. Taking just a few minutes to cross the continent, e-mail was much faster than the postal service, soon derisively called "snail mail." Besides being cheaper than a long-distance phone call, e-mail eliminated the need for both parties having to synchronize their activities, freeing them from their desks. E-mail also eliminated some of the problems associated with different time zones. Thus a New Yorker arriving at his or her office in the morning could e-mail a memo to a colleague in Los Angeles, even though the working day on the West Coast had not yet begun; the e-mail would be waiting in the recipient's electronic "in" tray at the start of the business day.

It was not just e-mail users who liked the new way of doing business; managers too had become enthusiastic about its economic benefits, especially for coordinating group activity. Meetings were still needed, but there was an expectation—perhaps seldom fulfilled—that there would be far fewer of them because team members could be kept in touch between meetings through group mailings. Managers claimed that activities that had once taken weeks could now be telescoped into days. By the mid-1980s Computer Supported Cooperative Working (CSCW), as such activities had come to be called, had become an established interdisciplinary research area that cut across the old subject boundaries of the computer and the social sciences.

Electronic mail was a completely new communications medium, however, and brought with it a range of social issues that fascinated organizational psychologists and social scientists. For example, the speed of communications encouraged knee-jerk rather than considered responses, thus increasing rather than decreasing the number of exchanges. Another problem was that the terse "telegraphese" of e-mail exchanges could easily cause offense to the uninitiated, when the same message conveyed over the phone would have been softened by the tone of voice and a more leisurely delivery. Gradually an unwritten "netiquette" emerged, enabling more civilized interactions to take place without losing too many of the benefits of e-mail.

Although some of the new computer networks of the 1970s were based on the technology developed in ARPANET, this was not true of all of them. The computer manufacturers, in particular, developed their own systems, such as IBM's Systems Network Architecture (SNA) and Digital Equipment Corporation's DECNET. By reworking a very old marketing strategy, the manufacturers were hoping to keep their customers locked into proprietary networks for which they would supply all the hardware and software. This was a shortsighted and mistaken strategy because the real benefits of networking came through *inter*networking—when the separate networks were connected and literally every computer user could talk to every other.

Fortunately, IPTO had been aware of this problem as early as 1973—not least so that electronic mail would be able to cross network boundaries. Very quickly, what was simply an idea, internetworking, was made concrete as the Internet. The job of IPTO at this stage was to establish "protocols" by which the networks could communicate. (A network protocol is simply the ritual electronic exchange that enables one network to talk to another—a kind of electronic Esperanto.) The system that ARPA devised was known as the transmission control protocol/Internet protocol—or simply TCP/IP—a mysterious abbreviation familiar to most experienced users of the Internet. Although the international communications committees were also trying to evolve internetworking standards at the same time, TCP/IP quickly established itself as the de facto standard (which it remains).

But it would be a full decade before significant numbers of networks were connected. In 1980 there were fewer than two hundred hosts on the Internet, and as late as 1984 there were still only a thousand. For the most part these networks served research organizations and science and engineering departments at universities—a predominantly technical community using conventional time-sharing computers. The Internet would become an important economic and social phenomenon only when it also reached the broad community of ordinary personal-computer users.

The World Wide Web

This broad community of users began to exert its influence in the late 1980s. In parallel with the development of the Internet, the personal computer was spreading across the educational and business communities and finding its way into American homes. By the late 1980s most professional computer users (though not home users) in the United States had access to the Internet, and the number of computers on the Internet began an explosive growth.

Once this much broader community of users started to use the Internet, people began to exchange not just e-mail but whole documents. In effect a new electronic publishing medium had been created. Like e-mail and news groups, this activity was largely unforeseen and unplanned. As a result there was no way to stop anyone from publishing anything and placing it on the net; soon there were millions of documents but no catalog and no way of finding what was useful. Sifting the grains of gold amid the tons of trash became such a frustrating activity that only the most dedicated computer users had the patience to attempt it. Bringing order to this universe of information now became a major research issue.

There were numerous parallel developments to create finding aids for information on the Internet. One of the first systems, "archie," was developed at McGill University in 1990. By combing though the Internet, archie created a directory of all the files available for downloading so that a user wanting a file did not need to know on what machine it was actually located. A more impressive system was the Wide Area Information Service (WAIS) developed by the Thinking Machines Corporation of Waltham, Massachusetts, the following year. WAIS enabled users to specify documents by using keywords (say, "smallpox" and "vaccine"), and it would then display all the available documents on the Internet that matched those criteria. The most popular early finding aid was "gopher," developed at the University of Minnesota. (A highly appropriate name, gopher is both colloquial for one who runs errands, as well as the university's mascot.) The system became operational in 1991 and was effectively a catalog of catalogs that enabled a user to drill down and examine the contents of gopher databases maintained by hundreds of different institutions.

All of these systems treated documents as individual entities, rather like books in a library. For example, when one had located a document about smallpox vaccine, say, it might tell you that its inventor was Edward Jenner; in order to discover more about Jenner, one needed to search again. The inventors of hypertext—Vannevar Bush in the 1940s and Englebart and Nelson in the 1960s—had envisaged a system that would enable one to informally skip from document to document. As it were, at the press of button, one could leap from "smallpox" to "Jenner" to "The Chantry" in Gloucestershire, England (the house where Jenner lived and now a museum to his memory). Hypertext was, in fact, a lively computer research topic throughout the 1980s, but what made it so potent for the Internet—ultimately giving rise to the Word Wide Web—was that it would make it unnecessary to locate documents in centralized directories. Instead, links would be stored in the documents themselves, and they would instantly whisk the reader to a related document. It was all very much as Vannevar Bush had envisioned the memex.

The World Wide Web was invented by Tim Berners-Lee. Its origins dated back to an early interest Berners-Lee had in hypertext in 1980, long before the Internet was widely known. Berners-Lee was born in London in 1955, the son of two mathematicians (who were themselves pioneers of early British computer programming). After graduating in physics from Oxford University in 1976, he worked as a software engineer in the UK before obtaining a six-month consulting post at CERN (Conseil Européen pour la Recherche Nucléaire), the international nuclear physics research laboratory in Geneva. While at CERN, Berners-Lee was assigned to develop software for a new particle accelerator, but in his spare time he developed a hobby program for a hypertext system that he called Enquire. (The program was named after a famous long-out-of-print Victorian household compendium called *Enquire Within Upon Everything*—his parents had owned a copy, and as a child he had enjoyed browsing through it.) Despite the similarity to Bush's vision, Berners-Lee had no firsthand knowledge of his work—rather Bush's ideas had simply become absorbed into the hypertext ideas that were in the air in the 1980s. There was nothing very special about Enquire: it was simply another experimental hypertext system like dozens of others. When he left CERN, it was effectively orphaned.

The personal-computer boom was in full swing when Berners-Lee returned to England, and he found gainful employment in the necessary, if rather mundane, business of developing software for dot-matrix printers. In September 1984, he returned to CERN as a permanent employee. In the years he had been away, computer networking had blossomed. CERN was in the process of linking all its computers together, and he was assigned to help in this activity. But it was not long before he dusted off his Enquire program and revived his interest in hypertext. Berners-Lee has described the World Wide Web as "the marriage of hypertext and the Internet." But it was not a whirlwind courtship; rather it was five years of peering though the fog as the technologies of hypertext and the Internet diffused and intertwined. It was not until 1989 that Berners-Lee and his Belgian collaborator Robert Cailliau got so far as to make a formal project proposal to CERN for the resources to create what they had grandiosely named the World Wide Web. (In fact, the CERN project turned out to be not much more than a fancy telephone directory. The Internet roller coaster did not wait while CERN dithered over funding.)

There were really two halves to the World Wide Web concept, known as the server and the client sides. The server would deliver hypertext documents (later know as Web pages) to a client computer, typically a personal computer or a workstation, that would display them on the user's screen. By the late 1980s hypertext was a well-established technology that had gone far beyond the academy into consumer products such as CD-ROM encyclopedias, although it had not yet crossed over to the Internet. By 1991 Berners-Lee and Cailliau were sufficiently confident of their vision to offer a paper on the World Wide Web to the Hypertext '91 conference in December in San Antonio, Texas. Their paper was turned down, but they managed to put on a halfhearted demonstration. Theirs was the only project that related to the Internet. Berners-Lee later reflected that when he returned to the conference in 1993 "every project on display would have something to do with the Web."

During that two-year period, the World Wide Web had taken off. It was a classic chicken-and-egg situation. Individuals needed Web "browsers" to read Web pages on their personal computers and workstations, while organizations needed to set up Web servers filled with interesting and relevant information to make the process worthwhile. Several innovative Web browsers came from universities, often developed by enthusiastic students. The Web browser developed at CERN had been a pedestrian affair for text-only hypertext documents, not hypermedia documents that were enriched with pictures, sounds, and video clips. The new generation of browsers provided the user-friendly point-and-click interface that people were starting to demand, as well as full-blown hypermedia features. First came the Erwise browser from the University of Helsinki in 1991, then Viola from the University of California at Berkeley a few months later, quickly followed by Midas, Samba, Lynx, Cello, and Mosaic for different computer platforms. Web servers needed software considerably more complex than a Web browser, although it was largely invisible to the average user. Here again, volunteer efforts, mainly from universities, produced serviceable solutions. The programs evolved rapidly as many individuals, communicating through the Internet, supplied bug fixes, program "patches," and other improvements. The most well-known server program to evolve by this process was known as "apache," a pun on "a patchy" server. In what was later known as the open-source movement, the Internet enabled collaborative software development to flourish to an extent previously unknown. And the fact that open-source software was free offered an entirely new model for software development to that of the conventional for-profit software company.

While all this was happening Berners-Lee had a useful index of how the Web was taking off—the number of "hits" on his original Web server at CERN. The number of hits grew from a hundred a day in 1991 to a thousand in 1992 to ten thousand in 1993. By 1994 there were several hundred publicly available Web servers, and the World Wide Web was rapidly overtaking the gopher system in popularity. This was partly because in spring 1993 the University of Minnesota had decided to assert ownership of the intellectual property in the gopher software and no doubt would eventually commercialize it. It was perhaps the first time that the commercialization of Internet software had surfaced, and there would forever after be an uneasy tension between the nascent open-source community, which promoted free software, and entrepreneurs who saw a business opportunity.

At all events, Berners-Lee was on the side of the angels when it came to commercial exploitation. He persuaded CERN to place Web technology in the public domain, for all to use free of charge for all time. In summer 1994 he moved from CERN to the Laboratory for Computer Science at MIT, where he would head the World Wide Web Consortium (W3C), a nonprofit organization to encourage the creation of Web standards though consensus.

Paradigm Shift

What started the extraordinary growth of the World Wide Web was the Mosaic browser. The first Web browsers mostly came from universities; they were hastily written by students, and it showed. The programs were difficult to install, buggy, and had an unfinished feel. The Mosaic browser from the National Center for Supercomputer Applications (NCSA) at the University of Illinois at Urbana-Champaign was the exception.

Developed by a twenty-two-year-old computer science undergraduate, Marc Andreessen, Mosaic was almost like shrink-wrapped software you could buy in a store. There was a version for the PC, another for the Macintosh, and another for the Unix workstations beloved of computer science departments. Mosaic was made available in November 1993, and immediately thousands—soon hundreds of thousands—of copies were downloaded. Mosaic made it very easy for owners of personal computers to get started surfing the Web. Of course they had to be enthusiasts, but no deep knowledge was needed.

In spring 1994 Andreessen received an invitation to meet with a California entrepreneur, Jim Clark. In the 1970s Clark had cofounded Silicon Graphics, a highly successful maker of Unix workstations. He had recently sold his share of the company and was on the lookout for a new start-up opportunity. The upshot of the meeting was that on 4 April 1994 the Mosaic Communications Corporation was incorporated to develop browsers and server software for the World Wide Web. The name of the corporation was changed to Netscape Communications a few months later, because the University of Illinois licensed the Mosaic name and software to another firm, Spyglass Inc.

Andreessen immediately hired some of his programming colleagues from the University of Illinois and got started cranking out code, doing for the second time what they had done once before. The results were very polished. In order to quickly establish the market for their browser, Clark and Andreessen decided to give it away free to noncommercial users. When they had grabbed the lion's share of the market, they could perhaps begin to charge for it. Server software and services were sold to corporations from the beginning, however. This established a common pattern for selling software for the Internet.

In December 1994, Netscape shipped version 1.0 of its browser and complementary server software. In this case "shipping" simply meant making the browser available on the Internet for users—who downloaded it by the millions. From that point on the rise of the Web was unstoppable: By mid-1995 the Web accounted for a quarter of all Internet traffic, more than any other activity.

In the meantime Microsoft, far and away the dominate force in personal computing, was seemingly oblivious to the rise of the Internet. Its on-line service, MSN, was due to be launched at the same time as the Windows 95 operating system in August 1995. A proprietary network from the pre-Internet world, MSN had passed the point of no return, but it would prove an embarrassment of mistiming. Microsoft covered its bets by licensing the Mosaic software from Spyglass and including a browser dubbed Internet Explorer with Windows 95, but it was a lackluster effort.

Microsoft was not the only organization frozen in the headlights of the Internet juggernaut. A paradigm shift was taking place—a rapid change from one dominant technology to another. It was a transition from closed proprietary networks to the open world of the Internet. Consumers also had a difficult choice to make. They could either subscribe to one of the existing consumer networks—AOL, CompuServe, Prodigy, or MSN—or go with a new type of supplier called an Internet service provider (ISP). The choice was between the mature, user-friendly, safe, and content-rich world of the consumer network, or the wild frontier of the World Wide Web. The ISPs did not provide much in the way of content—there was plenty of that on the World Wide Web, and it was rapidly growing. The ISP was more like a telephone service—it gave you a connection, but then you were on your own. The ISP customer needed to be a little more computer savvy to master the relatively complex software and had to be wary to avoid the less savory aspects of the Internet (or was free to enjoy them).

The existing consumer networks had a huge technical and cultural challenge to respond to the World Wide Web. AOL licensed Microsoft's Internet Explorer for use in its access software. This gave its subscribers the best of both worlds—the advantages of the existing service, plus a window into the World Wide Web. CompuServe did much the same, though it was eventually acquired by AOL in 1998. The other consumer networks did not manage to emulate AOL's masterly segue. Prodigy's owners, Sears and IBM, decided to sell the network in 1996, and it thereafter faded from sight. Microsoft decided to cut its losses on MSN.

In December 1995 Gates announced that Microsoft would "embrace and extend" the Internet; it "was willing to sacrifice one child (MSN) to promote a more important one (Internet Explorer)." Thereafter, MSN was not much more than an up-market ISP—it was no threat to AOL.

Whereas in the 1980s the operating system had been the most keenly fought territory in personal computing, in the 1990s it was the Web browser. In 1995 Netscape had seemed unstoppable. In the summer—when the firm was just eighteen months old—its initial public offering netted $2.2 billion, making Marc Andreessen surely the youngest self-made multimillionaire in history. By the year's end its browser had been downloaded 15 million times, and it enjoyed over 70 percent of the market. But Microsoft did not intend to concede the browser market to Netscape.

During the next two years Microsoft and Netscape battled for browser supremacy, each producing browser upgrades every few months. By January 1998, with a reported investment of $100 million a year, version 4.0 of Microsoft's Internet Explorer achieved technical parity with Netscape. Because Internet Explorer was bundled at no extra cost with the new Windows 98 operating system, it became the most commonly used browser on PCs—perhaps not through choice, but by the path of least resistance. A version of Internet Explorer was also made available free of charge for the Macintosh computer.

Distributing Internet Explorer to consumers at no cost completely undermined the business plans of Netscape. How could it sell a browser when Microsoft was giving its away for nothing? Microsoft had fought hard, perhaps too hard, for its browser monopoly. Its alleged practices of tying, bundling, and coercion provoked the U.S. Department of Justice to file an antitrust suit in May 1998. Like all antitrust suits, this one proceeded slowly—especially for an industry that spoke of "Internet time" in the way that canine enthusiasts spoke of dog years. Microsoft's self-serving and arrogant attitude in court did nothing to endear it to consumers or the press, though it did little to arrest its progress. In the five years between 1995, when Microsoft was first wrong-footed by the rise of the Internet, and 2000, when Judge Thomas Penfield Jackson produced his verdict, Microsoft's revenues nearly quadrupled, from $6 billion to $23 billion, and it staff numbers doubled, from 18,000 to 39,000. Judge Jackson directed that Microsoft be broken up—a classic antitrust remedy for the alleged wrongdoings of a monopolist—although less drastic remedies were agreed upon in the final settlement. Quite how history will remember Microsoft's brushes with the Department of Justice one cannot tell—but if IBM's major three antitrust suits with the government (1932–36, 1952–56, and 1969–82) are any guide, they will seem minor skirmishes alongside the great sweep of technological progress. As discussed in chapter 6, what brought IBM low in the 1980s was not an antitrust suit but a failure to manage a shift in the technological landscape.

Dot Commerce

During the second half of the 1990s, the Internet became not so much an information revolution as a commercial revolution—users were able to purchase the full range of goods and services that society could offer. In 1990, however, the Internet was largely owned by agencies of the U.S. government, and the political establishment did not permit the use of publicly owned assets for private profit. Hence, privatization of the Internet was an essential precursor for electronic commerce to flourish.

Even before privatization had become an issue, it had been necessary to separate the civilian and military functions of the Internet. Since its inception in 1969, the ARPANET had been funded by the Advanced Research Projects Agency. In 1983 the military network was hived off as Milnet, and ARPANET became the exclusive domain of ARPA's research community. Once the military constraints were removed, the network flourished—by 1985 some 2,000 computers had access to the Internet. To broaden access to the U.S. academic community as a whole, beyond the exclusive circle of ARPA's research community, the National Science Foundation had created another network, NSFnet. In a way that proved characteristic of the development of the Internet, NSFnet rapidly overtook ARPANET in size and importance, eventually becoming the backbone of the entire Internet. By 1987 some 30,000 computers had access to the Internet—mostly from the U.S. academic and research communities. At the same time, other public and private networks attached themselves, such as Usenet, FidoNet (created by amateur "bulletin board" operators), and IBM's Bitnet. Who paid for their access was moot, because the NSF did not have any formal charging mechanisms. It was a wonderful example of the occasional importance of turning a blind eye. If any bureaucrat had looked closely at the finances, the whole project for integrating the world's computer networks would likely have been paralyzed. As more and more commercial networks came on board, additional infrastructure was added and they evolved their own labyrinthine mechanisms for allocating the costs. By 1995, the commercially owned parts of the Internet far exceeded those owned by the government. On 30 April 1995, the old NSFnet backbone was shut down, ending altogether U.S. government ownership of the Internet's infrastructure.

The explosive growth of the Internet was in large part due to its informal, decentralized structure—anyone was free to join in. However, the Internet could not function as a commercial entity in a wholly unregulated way or chaos and lawlessness would ensue. The minimal, light-touch regulation that the Internet pioneers evolved was one of its most impressive features. A good example of this is the domain name system. Domain names—such as Amazon.com, whitehouse.gov, and princeton.edu—soon became almost as familiar as telephone numbers.

In the mid-1980s the Internet community adopted the domain name system devised by Paul Mockapetris of the Information Sciences Institute at the University of Southern California. The system would decentralize the allocation of thousands (and eventually millions) of domain names. The process started with the creation of six top-level domains, each denoted by a three-letter suffix: *com* for commercial organizations, *edu* for educational institutions, *net* for network operators, *mil* for military, *gov* for government, and *org* for all other organizations. Six registration authorities were created to allocate names within each of the top-level domains. Once an organization had been given a unique domain name, it was free to subdivide it within the organization by adding prefixes as needed, without the necessity of permission from any external authority. For example, the Department of Computer Science at Princeton used the domain name cs.princeton.edu; even though other universities also used the *cs* prefix for their computer science departments, there was no conflict because the domain name taken as a whole was guaranteed to be unique.

Outside the United States, countries were allocated a two-letter suffix—*uk* for the United Kingdom, *ch* for Switzerland, and so on. Each country was then free to create its own second-level domain names. In Britain, for example, and for no very good reason, *ac* was used for educational (i.e., academic) institutions, *co* for commercial organizations, and so on. Thus the Computer Laboratory at Cambridge had the domain name cl.cam.ac.uk, while bbc.co.uk needs no explanation. The United States is the only country that does not use a country suffix; in this regard it is exercising a privilege rather like that of Great Britain, which invented the postage stamp in 1841 and is the only country whose stamps do not bear the name of the issuing nation.

If one examines the early predictions of how the Internet would evolve, they are very much extrapolations of H. G. Wells's World Brain or Vannevar Bush's memex. In 1937 Wells wrote of his World Brain: "The time is close at hand when any student, in any part of the world, will be able to sit with his projector in his own study at his or her own convenience to examine *any* book, *any* document, in an exact replica." In 1996 (when the first edition of this book went to press) this seemed a reasonable prediction, but one that might be twenty or thirty years in the making. Although we see no reason to revise the timeline, the progress made has nonetheless been astonishing. Moreover, the information resources on the Internet go far beyond books and documents but embrace audio, video, and multimedia—which Wells did not predict. A large fraction of the world's current printed output, especially research journals and news magazines, is now available on-line. Indeed, for university and industrial researchers most academic journals are available over the Internet, and visits to libraries are made with ever decreasing frequency. For the historian or the humanities scholar the situation is somewhat different, because the vast scholarship of the millennia has not yet become available on-line, and for them there is still no substitute for a visit to the world's great libraries.

Extraordinary as the progress of the Internet has been as an information resource, even more extraordinary has been the rise of electronic commerce—a phenomenon largely unpredicted as late as the mid-1990s.

Some of the early business successes were almost accidental. For example, Yahoo!, one of the most visited websites on the Internet, began in the days of the Mosaic browser as a simple listing service ("Jerry's Guide to the World Wide Web") created by David Filo and Jerry Yang, two computer science graduate students at Stanford University. By late 1993, the site listed a modest 200 websites, though at the time that was a significant fraction of the world's websites. During 1994, however, the Web experienced explosive growth. As new sites came on-line daily, Filo and Yang sifted, sorted, and indexed them. Toward the end of 1994 Yahoo! experienced its first million-hit day—representing perhaps 100,000 users. The following spring Filo and Yang secured venture capital, moved into offices in Mountain View, California, and began to hire staff to surf the Web to maintain and expand the index. It was very much as Wells had envisaged for the World Brain when he wrote of "a great number of workers . . . engaged in perfecting this index of human knowledge and keeping it up to date." Yahoo! was not without competition: Lycos, Excite, and a dozen others had come up with the same concept, and listing services became one of the first established categories of the Web. One question remained: How to pay for the service? The choices included subscriptions, sponsorship, commissions, and advertising. As with early broadcasting, advertising was the obvious choice.

Another early Web success was selling by mail order. Jeff Bezos's Amazon.com established many of the early practices (including the incorporation of "com" in the firm's name). Bezos, a Princeton graduate in electrical engineering and computer science, had enjoyed a brief career in financial services before deciding to set up a retail operation on the Web. After some deliberation, he lighted on bookselling as being the most promising opportunity—there were relatively few major players in bookselling, and a virtual bookstore could offer an inventory far larger than any brick-and-mortar establishment. Launched in July 1995, Amazon.com was not the first bookshop on the Web, but by offering deep discounts on popular titles and through brilliant management of the new medium—for example, by inviting readers to submit personal book reviews—it quickly dominated its niche. Amazon.com became a virtual community. Customers returned again and again—sometimes just to browse, sometimes to post reviews, and sometimes to buy.

The eBay auction website was another operation that succeeded largely through mastery of the new medium. The auction concept was developed as an experimental free website, AuctionWeb, by Pierre Omidyer, a Silicon Valley entrepreneur, in late 1995. The site proved very popular, and by 1996 Omidyer was able to begin charging selling commissions, after which "he left his day job and changed the site's name to eBay." In early 1998 professional management was brought in, in the form of Meg Whitman, a senior executive from the toy maker Hasbro. Over the next few years, a worldwide virtual community of buyers and sellers evolved. By 2003, eBay had 30 million users worldwide and $20 billion in sales, and it was said that 150,000 entrepreneurs made their living as eBay traders. As an example of the transforming power of the World Wide Web on commerce, a global flea market is as unlikely a concept as one could have imagined. It speaks volumes for the unpredictably of the Web and the surprises that are yet in store.

For all of their dramatic growth, profits from the commercial Web ventures such as Yahoo!, Amazon.com, and eBay remained elusive even as their stock prices soared. Inevitably, when Internet euphoria subsided, stock prices plunged—in spring 2000 many prominent e-commerce operations lost as much as 80 percent of their value. It had happened before: In the late 1960s software and computer utility stocks suffered a similar collapse, from which it took several years to recover. But there was never any doubt that recovery would come, and today the software crash of 1969 is but a distant memory, and only for those who lived through it. The same will prove true for the dot-com crash of 2000—although this may be small comfort to individuals nursing stock portfolios bought at the top of the market.

Because of its global reach and the absence of national boundaries, the Internet has proved extremely difficult to regulate. For example, by the year 2000 the problem of unsolicited e-mail—"spam"—had become endemic and was

endangering e-mail's efficacy, not to mention recipients' sensibilities and toler-
ance. However, unsolicited e-mail was not illegal in the United States, any
more than was postal junk mail, even though its disruptive effect was far
worse. In 2003, antispam laws were enacted in several countries; they are a beginning, but they are far from being wholly effective. For example, unscrupulous operators can simply move their operations offshore to countries where
Internet regulation is not yet on the agenda. International law is capable of re-
solving these issues, but it will take time—perhaps many years—to achieve an
international consensus.

Yet, within developed nations, there have been effective remedies against
some disruptive or illegal operations. A much-publicized example was the action taken against Napster by the Record Industry Association of America
(RIAA). Napster was the brilliant invention of a seventeen-year-old college
freshman, Shawn Fanning, that enabled users to swap music files over the Internet. A business was incorporated as Napster Inc. in July 1999. Fanning's invention did not require a central computer to store music files but merely facilitated
the exchange of files by users. The direct copying of music (other than "fair
use") is of course illegal, but Napster believed that merely facilitating file sharing
did not amount to copyright infringement. RIAA filed suit within a few months
of Napster's incorporation. This was by no means a black-and-white case. To
many commentators RIAA appeared to be breaking a butterfly on a wheel, while
to others Napster seemed to threaten the very existence of the media industries
that depended on copyright protection. Ultimately Napster was closed down.
One day the world will cease to buy music on physical media, and the Napster-
RIAA dispute will be seen to be but one skirmish on the path from the old world
to the new.

Regulation of the Internet is exercising policy makers in all developed countries, but progress is slow. Perhaps the history of broadcasting provides the
most instructive analogy. In the early days of radio, shortly after World War I,
there was little or no regulation—almost anyone could set up a radio station
wherever they wanted, provided it did not endanger military or shipping uses
of radio. The result was such chaos that it threatened the very existence of radio
as an entertainment medium. In the United States, legislation in the Radio Act
of 1927 provided some immediate remedies and eventually led to the formation
of the Federal Communications Commission in 1934, the body that still over-
sees American broadcasting. Other countries evolved their own local regulatory
arrangements (for example, in the UK a national monopoly, the BBC, was created in 1926 to undertake all broadcasting). With the rise of international

short-wave broadcasting in the 1930s and 1940s, the International Telecommunications Union played a similar (though less effective) role. One way or another, the regulating authorities resolved conflicting private and public
interests, mitigated cross-station interference, and guaranteed the decency of
public broadcasting. As we write, the Internet is as unregulated as radio was in
the early twentieth century—it is open to abuse by any teenage virus writer, in-
discriminate "spam house," cybersquatter, copyright infringer, or unscrupulous
pornographer. It is far from clear how these abuses can be resolved, as the devil
will be in the details; yet we are confident that in time they will be resolved—
the Internet is simply too important for its continued existence to be imperiled
by an antisocial and lawless minority.

:: NOTES

A NOTE ON SOURCES

Between the extremes of giving a popular treatment with no citations at all and full scholarly annotation, we have decided on a middle course in which we have attempted to draw a line between pedantry and adequacy—as Kenneth Galbraith eloquently put it. In the notes to every chapter, we have given a short literature review in which we have identified not more than a dozen reliable secondary sources. Where possible, we have used the monographic literature, falling back on the less accessible periodical literature only where the former falls short. The reader can assume, unless explicitly indicated, that every assertion made in the text can be readily found in this secondary literature. This has enabled us to restrict citations to just two types: direct quotations and information that does not appear in the monographs already cited.

Full details are given in the bibliography. The reader should note that its purpose is largely administrative—to put all our sources in one place. It is not intended as a reading list. If a reader should want to study any topic more deeply, we recommend beginning with one of the books cited in the literature review in the notes to each chapter.

CHAPTER 1. WHEN COMPUTERS WERE PEOPLE

A modern account of table making is *The History of Mathematical Tables* (2003), edited by Martin Campbell-Kelly et al. This contributed volume includes chapters on Napier's logarithms, de Prony's *tables du cadastre*, Babbage's Difference Engines, and other episodes in the history of tables. The "tables crisis" and Babbage's calculating engines are described in Doron Swade's very readable *The Difference Engine: Charles Babbage and the Quest to Build the First Computer* (2001) and his earlier, beautifully illustrated *Charles Babbage and His Calculating Engines* (1991). Swade writes with great authority and passion, as he was the curator responsible for building Babbage's Difference Engine No. 2 to celebrate the bicentennial of Babbage's birth in 1991. Detailed accounts of Victorian data processing are given in Martin Campbell-Kelly's articles on the Prudential Assurance Company, the Railway Clearing House, the Post Office Savings Bank, and the British Census (1992, 1994, 1996, and 1998) and Jon Agar's *The Government Machine* (2003). The early American scene is described in Patricia Cline Cohen's *A Calculating People* (1982).

The political, administrative, and data-processing background to the U.S. Census is given in Margo Anderson's *The American Census* (1988). The most evocative description of the 1890 U.S. Census is given in a contemporary account by Thomas C. Martin, "Counting a Nation by Electricity" (1891); a more prosaic account is given in Leon Truesdell's *The Development of Punch Card Tabulation in the Bureau of the Census, 1890–1940* (1965). The standard biography of Hollerith is Geoffrey D. Austrian's *Herman Hollerith: Forgotten Giant of Information Processing* (1982). The early office-systems movement is described in JoAnne Yates's *Control Through Communication* (1989).

Page

4	**"fell on evil days"**: Comrie 1933, p. 2.
5	**"conceived all of a sudden"**: Quoted in Grattan-Guinness 1990, p. 179.
6	**"one of the most hated symbols"**: Ibid.
7	**an open letter to . . . Sir Humphrey Davy**: Babbage 1822a.
5	**"render the navigator liable"**: Lardner 1834.
6	**"An undetected error"**: Quoted in Swade 1991, p. 2.
8	**"navigational tables were full of errors"**: Hyman 1982, p. 49.
9	**a letter . . . to the Duke of Wellington**: Babbage 1834.
9	**British and U.S. clearing houses**: For descriptions, see Cannon 1910.
10	**Babbage wrote to Lubbock**: Quoted in Alborn 1994, p. 10.
10	**"In a large room in Lombard Street"**: Babbage 1835, p. 89.
13	**"every town of importance"**: Anon. 1920, p. 3.
14	**"It is a cheerful scene of orderly industry"**: Anon. 1874, p. 506.
14	**Bureau of the Census**: The administrative organization is described in Anderson 1988. Head count and page count statistics appear on p. 242.
15	**"tally system"**: Described in Truesdell 1965, chap. 1.
15	**"The only wonder . . ."**: Martin 1891, p. 521.
16	**"I was traveling in the West"**: Austrian 1982, p. 15.
16	**Porter . . . would later become chairman of the British Tabulating Machine Company**: Campbell-Kelly 1989, pp. 16–17.
17	**"rounded up . . . every square foot"**: Austrian 1982, p. 59.
18	**"feel their nation's muscle"**: Ibid., p. 58.
17	**"The statement by Mr. Porter"**: Martin 1891, p. 522.
18	**The press loved the story**: The quotations appear in Austrian 1982, pp. 76 and 86.

18   **"vastly interesting," "augurs well," and "women show":** Martin 1891, p. 528.
19   **"In other words":** Ibid., p. 525.
19   **"for all the world like that of sleighing":** Ibid., p. 522.
19   **"the apparatus works as unerringly":** Ibid., p. 525.
19   **"Herman Hollerith used to visit":** Quoted in Campbell-Kelly 1989, p. 13.
20   **By contrast, the American Prudential:** See May and Oursler 1950; Yates 1993.
21   **"Now, administration without records":** Quoted in Yates 1989, p. 12.

CHAPTER 2. THE MECHANICAL OFFICE

The best modern account of the office-machine industry is James W. Cortada's *Before the Computer* (1993).

Good histories of the typewriter are Wilfred Beeching's *Century of the Typewriter* (1974) and Bruce Bliven's *Wonderful Writing Machine* (1954). George Engler's *The Typewriter Industry* (1969) contains a fine analysis of the industry not available elsewhere. The only analytical account of filing systems we are aware of is JoAnne Yates's excellent article "From Press Book and Pigeon Hole to Vertical Filing" (1982). There are no reliable in-depth histories of Remington Rand, Felt & Tarrant, or Burroughs, but Cortada's *Before the Computer* covers most of the ground.

Useful histories of National Cash Register are Isaac F. Marcosson's *Wherever Men Trade* (1945) and Stanley C. Allyn's *My Half Century with NCR* (1967). In 1984 NCR produced an excellent, though difficult to find, company history in four parts (NCR 1984). The best biography of John Patterson is Samuel Crowther's *John H. Patterson, Pioneer in Industrial Welfare* (1923). NCR's R&D operations are described in Stuart W. Leslie's *Boss Kettering* (1983).

Histories of IBM abound. The best accounts of the early period are Saul Engelbourg's *International Business Machines: A Business History* (1954 and 1976), Emerson W. Pugh's *Building IBM* (1995), and Robert Sobel's *Colossus in Transition* (1981). The official biography of Thomas J. Watson is Thomas Beldens and Marva Beldens's *The Lengthening Shadow* (1962); it is not particularly hagiographic, but William Rogers's *THINK: A Biography of the Watsons and IBM* (1969) is a useful counterbalance.

Page
25   **"let them in on the ground floor":** Quoted in Bliven 1954, p. 48.
25   **the QWERTY keyboard layout that is still with us:** See David 1986.
25   **"typewriter was the most complex mechanism mass produced by American industry":** Hoke 1990, p. 133.
25–26 **"reporters, lawyers, editors, authors, and clergymen":** Cortada 1993, p. 16.

26   "Gentlemen: Please do not use my name": Quoted in Bliven 1954, p. 62.
27   By 1900 the U.S. Census recorded 112,000 typists: Davies 1982, pp. 178–79.
27   "a woman's place is at the typewriter": Ibid.
28   James Rand Sr.: Hessen 1973.
30   labeled 1 through 9: Because zeros did not need to be added into the result, no zero keys were provided.
32   "One Built for Every Line of Business": From an advertisement reproduced in Turck 1921, p. 142.
33   "business practices and marketing methods": Cortada 1993, p. 66.
33   "the only thing he knew anything about was coal": Quoted in Crowther 1923, p. 72.
34   "turn off the worn, unexciting highway": Allyn 1967, p. 54.
35   "You insure your life": Quoted in Cortada 1993, p. 69.
35   "the first canned sales approach": Belden and Belden 1962, p. 18.
35   "The school taught the primer": Crowther 1923, p. 156.
36   "the best way to kill a dog": Quoted in Rogers 1969, p. 34.
36   "ever onward . . .": Anon. 1932, p. 37.
38   "The system is used in factories": Quoted in Campbell-Kelly 1989, p. 31.
40   "Everywhere . . . there will be IBM machines": Ibid.
40   "[L]ike a gifted plagiarist": Ibid.
41   "a type that might be called 'refill' businesses": Ibid., p. 39.
42   "Watson had one of the supreme moments in his career": Belden and Belden 1962, p. 114.
42   "the most lucrative of all IBM's mechanical glories": Anon. 1940, p. 126.
43   "the world's biggest bookkeeping job": Quoted in Eames 1973, p. 109.
44   "Most enterprises, when they would dream": Anon. 1940, p. 36.

CHAPTER 3. BABBAGE'S DREAM COMES TRUE

Michael R. Williams, *A History of Computing Technology* (1997) and William Aspray (ed.), *Computing Before Computers* (1990) both give a good general overview of the developments described in this chapter. Alan Bromley's article "Charles Babbage's Analytical Engine, 1838" (1982) is the most complete (and most technically demanding) account of Babbage's great computing machine, whereas Bruce Collier and James MacLachlan's *Charles Babbage and the Engines of Perfection* (1998) is a very accessible account aimed at youngsters but perfect for mature readers too. Bromley's contributed chapter "Analog Computing" in Aspray's *Computing Before Computers* is the best historical overview of nondigital computing. Frederik Nebeker's *Calculating the Weather* (1995) is the best modern account of numerical meteorology. Comrie's Scientific Computing Service is described in Mary Croarken's *Early Scientific Computing in Britain* (1990). A fascinating example of human computing between the wars is described in David Grier's article "The Math Tables Project: The Reluctant Start of the Computing Era" (1998), while Paul Ceruzzi has written a fine account of wartime manual of computing activities in his article "When Computers Were Human" (1991).

This chapter tracks the development of automatic computing through the contributions of a number of pivotal figures: Charles Babbage, L. J. Comrie, Lewis Fry Richardson, Lord Kelvin, Vannevar Bush, and Howard Aiken. Of these, all but Comrie have standard biographies. They are Anthony Hyman's *Charles Babbage: Pioneer of the Computer* (1984), Oliver Ashford's *Prophet—or Professor? The Life and Work of Lewis Fry Richardson* (1985), Crosbie Smith and Norton Wise's *Energy and Empire: A Biographical Study of Lord Kelvin* (1989), G. Pascal Zachary's *Endless Frontier: Vannevar Bush, Engineer of the American Century* (1997), and I. Bernard Cohen's *Howard Aiken: Portrait of a Computer Pioneer* (1999). A vignette of Comrie is given by Mary Croarken in her article "L. J. Comrie: A Forgotten Figure in the History of Numerical Computation" (2000), while his milieu is well described in her *Early Scientific Computing in Britain* (1990).

Page

45   "The black mark earned": Comrie 1946, p. 567.

46   "eating its own tail": Quoted in Bromley 1990a, p. 75.

47   "a totally new engine" and "you may have fairly": Babbage 1834, p. 6.

48   "The distinctive characteristic": Lovelace 1843, p. 121.

49   "dear and much admired Interpreter": Toole 1992, pp. 236–39.

49   "worthless": Quoted in Campbell-Kelly's introduction to Babbage 1994, p. 28.

49   "Propose to an Englishman": Babbage 1852, p. 41.

50   Scheutz Difference Engine: See Lindgren 1990.

52   analog computing: For an excellent short history of analog computing, including orreries and Kelvin's tide predictor, see Bromley 1990b.

52–53   Similar techniques were used in land-reclamation planning in the Netherlands: Van den Ende 1992 and 1994.

53   "It consisted of an instrument box": Bush 1970, pp. 55–56.

54   a slightly blurred photograph: The photograph is reproduced in Zachary 1997, facing p. 248 and Wildes and Lindgren 1986, p. 86.

54   "So some young chaps": Bush 1970, p. i6i.

55, 56   "heap of hay in a cold rest billet" and following quotations: Richardson 1922, pp. 219–20.

57   £100 million: Ashford 1985, p. 89.

59   "Girls Do World's Hardest Sums": Press cutting from *Illustrated*, 10 January 1942, in NAHC / SCS /A1, National Archive for the History of Computing, Manchester University.

59   **"vital to the entire research effort":** Ceruzzi 1991, p. 239.

60   **Eckert and Columbia University were to play a key role in IBM's transition to computing:** See Brennan 1971.

60   **"rather limited enthusiasm . . .":** Cohen 1988, p. 179.

60   **Chase was a very well-known and respected figure:** See I. B. Cohen's introduction to Chase 1980.

60   **"in spite of the large anticipated costs":** Pugh 1995, p. 73.

61   **"couldn't see why in the world":** Cohen 1988, p. 179.

61   **"If, unwarned by my example":** Babbage 1994, p. 338, and Babbage 1864, p. 450.

62   **"felt that Babbage was addressing him":** Bernstein 1981, p. 62.

62   **This proposal is the earliest surviving document:** Aiken 1937.

62   **"fully automatic in its operation" and "control to route the flow of numbers":** Ibid., p. 201.

62   **"Automatic Computing Plant":** Bashe et al. 1986, p. 26.

63   **"a nineteenth-century New England textile mill":** Ceruzzi's introduction to 1985 reprint of Aiken et al. 1946, p. xxv.

63   **"a roomful of ladies knitting":** Bernstein 1981, p. 64.

63   **"makin' numbers":** Hopper 1981, p. 286.

63   **"I guess I'm the only man in the world":** Cohen's foreword to the 1985 reprint of Aiken et al. 1946, p. xiii.

64   **Norman Bel Geddes:** Ibid.

64   **"Few events in Watson's life" and following quotations:** Belden and Belden 1962, pp. 260–61.

64   **"Harvard's Robot Super-Brain" and "Robot Mathematician Knows All the Answers":** Quoted in Eames and Eames 1973, p. 123, and Ceruzzi's introduction to the 1985 reprint of Aiken et al. 1946, p. xxxi.

64   **"One notes with astonishment":** Comrie 1946, p. 517.

65   **"Half a century may probably elapse":** Babbage 1994, p. 338, and Babbage 1864, p. 450.

CHAPTER 4. INVENTING THE COMPUTER

The best scholarly account of the Moore School developments is contained in Nancy Stern's *From ENIAC to UNIVAC: An Appraisal of the Eckert-Mauchly Computers* (1981). Herman G. Goldstine's *The Computer: From Pascal to von Neumann* (1972) contains an excellent firsthand account of the author's involvement in the development of the ENIAC and EDVAC. A "view from below" is given in Herman Lukoff's *From Dits to Bits* (1979); the author was a junior engineer on the ENIAC and later became a senior engineer in the UNIVAC Division of Sperry Rand. Scott

McCartney's *ENIAC* (1999) is an engaging read that contains biographical material on Eckert and Mauchly not found elsewhere.

There are two full-length books describing Atanasoff's contributions to computing: Alice and Arthur Burks's *The First Electronic Computer: The Atanasoff Story* (1988) and Clark R. Mollenhoff's *Atanasoff: Forgotten Father of the Computer* (1988).

The controversy over the invention of the stored-program concept is discussed at length in both the Stern and the Goldstine books, with some bias toward Eckert-Mauchly and von Neumann, respectively. A more objective account is contained in William Aspray's *John von Neumann and the Origins of Modern Computing* (1990).

The background to the Moore School Lectures is given in Martin Campbell-Kelly and Michael R. Williams's *The Moore School Lectures* (1985). The best accounts of computing at Manchester University and Cambridge University are contained in, respectively, Simon H. Lavington's *Manchester University Computers* (1976) and Maurice V. Wilkes's *Memoirs of a Computer Pioneer* (1988).

Page

70   "coordinate, supervise, and conduct scientific research": Quoted in Baxter 1946, p. 14.

70   "most of the worthwhile programs of NDRC" and "When once a project got batted into [a] form": Bush 1970, p. 48.

72   "Mrs. John W. Mauchly, a professor's wife": Lukoff 1979, p. 21.

72   "on the very few occasions we were allowed to see Annie": Ibid., p. 18.

74   "forgotten father of the computer": Mollenhoff 1988.

74   "no ideas whatsoever": Stern 1981, p. 8.

74   "undoubtedly the best electronic engineer in the Moore School": Goldstine 1972, p. 149.

75   "to work about ten times faster": Eckert 1976, p. 8.

75   The MTI device worked as follows: See Ridenour 1947, chap. 16.

76   "*at least* several hours": Mauchly 1942.

76   "In addition to a staff of 176 computers": Goldstine 1972, pp. 164–66.

77   "at best unenthusiastic and, at worst hostile": Stern 1981, p. 21.

77   "the consummate engineer" and "the visionary": Ibid., p. 12.

77   "Eckert's standards were the highest": Goldstine 1972, p. 154.

79   "I was waiting for a train to Philadelphia": Ibid., p. 182.

80   "Of course, this *was* von Neumann's first query": Ibid.

81   "great new technological invention": Ibid., p. 190.

84   "entirely without publicity and attendance by invitation only": Stern 1981, pp. 219–20.

86   "two others connected with the British Post Office": Goldstine 1972, p. 217.

88   **"took us by the hand"**: Williams 1975, p. 328.

88   **"In early trials it was a dance of death"**: Ibid., p. 330.

89   **"In the middle of May 1946 I had a visit from L. J. Comrie"**: Wilkes 1985, p. 108.

89   **"beginning to despair"**: Ibid., p. 116.

89   **"not going to lose very much in consequence of having arrived late"**: Ibid., p. 119.

89   **"basic principles of the stored program computer were easily grasped"**: Ibid., p. 120.

90   **"computer of modest dimensions"**: Ibid., p. 127.

91   **"I was rung up one day by a genial benefactor"**: Ibid., pp. 130–31.

CHAPTER 5. THE COMPUTER BECOMES A BUSINESS MACHINE

The most detailed account of the Eckert-Mauchly computer business is contained in Nancy Stern's *From ENIAC to UNIVAC: An Appraisal of the Eckert-Mauchly Computers* (1981), while Herman Lukoff's *From Dits to Bits* (1979) contains a great deal of atmospheric detail of the building of the UNIVAC.

The best overall account of IBM's computer developments is Emerson Pugh's *Building IBM* (1995). Technical developments are described in Charles Bashe et al.'s *IBM's Early Computers* (1986). Thomas J. Watson's autobiography *Father and Son & Co.* (1990) is a valuable, if somewhat revisionist, view from the top of IBM.

The many players in the computer industry in the 1950s are detailed in Paul Ceruzzi's *History of Modern Computing* (2003), James W. Cortada's *The Computer in the United States: From Laboratory to Market, 1930–1960* (1993), Katherine Fishman's *The Computer Establishment* (1981), and Franklin M. Fisher et al.'s *IBM and the U.S. Data Processing Industry* (1983).

Page

96   **"Eckert and Mauchly were more than optimistic"**: Stern 1981, p. 106.

99   **"difficult . . . to believe that we would ever outgrow this facility"**: Lukoff 1979, p. 81.

99   **"but very poorly"**: Quoted in Stern 1981, p. 124.

100   **"evolution not revolution"**: Campbell-Kelly 1989, p. 106.

100   **"find the most outstanding Professor on electronics and get him for IBM"**: Quoted in Pugh 1995, p. 122.

101   **"speed and flexibility of operation"**: Ibid., p. 131.

102   **"This machine was completed and put into operation by IBM"**: Bowden 1953, pp. 174–75.

102   **"[p]eople who just want to look at the calculator"** and **"The principal cerebral parts of the machine"**: Anon. 1950.

104   "primarily patriotic" and "seized on the idea as an opportunity": Pugh 1995, p. 170.

105   "I was curious about Mauchly, whom I'd never met" and "Mauchly never said a word": Watson 1990, pp. 198–99.

105   "so desperate that they were ready to consider the first reasonable offer": Stern 1981, p. 148.

106   "first, one bay was set up": Lukoff 1979, p. 96.

107   "Each day he worked a little later" and "Pres's nervous energy was so great": Ibid., p. 106.

107   "John was always a great morale booster": Ibid., p. 75.

107   "shorts and undershirts": Ibid., pp. 99–100.

107   "He had a half dozen soda bottles": Ibid., pp. 99–100.

108   "IT'S AWFULLY EARLY, BUT I'LL GO OUT ON A LIMB": Reproduced in ibid., p. 130.

108   "Our election officials . . .": Ibid., pp. 130–31.

109   "I thought . . . here we are trying to build Defense Calculators": Watson 1990, p. 227.

110   "People at IBM invented the term 'panic mode'": Ibid., p. 228.

111   "The longest-haired computer theorists": Anon. 1952, p. 114.

113   "While our giant, million-dollar 700 series got the publicity": Watson 1990, p. 244.

114   "became an important factor in making NCR's . . . computer development possible": Allyn 1967, p. 161.

115   "represented courage and foresight": Quoted in Fisher et al. 1983, p. 81.

CHAPTER 6. THE MATURING OF THE MAINFRAME: THE RISE AND FALL OF IBM

The competitive environment of the computer industry between 1960 and 1975 is described in Gerald W. Brock's *The U.S. Computer Industry* (1975). Other useful sources are Kenneth Flamm's monographs *Targeting the Computer: Government Support and International Competition* (1987) and *Creating the Computer: Government, Industry, and High Technology* (1988).

A history of the IBM 1401 is given in Charles J. Bashe et al.'s *IBM's Early Computers* (1986). The technical development of the IBM 360 and 370 computers and the abortive Future System is described in Emerson W. Pugh et al.'s *IBM's 360 and Early 370 Systems* (1991). Tom Wise's *Fortune* articles "I.B.M.'s $5,000,000,000 Gamble" and "The Rocky Road to the Market Place" (1966a and 1966b) remain the best external account of the System/360 program. A good internal account is contained in Thomas DeLamarter's *Big Blue* (1986). The best analysis of IBM's post-1990 demise is given in Steven W. Usselman's prize-winning essay "IBM and Its Imitators" (1993).

Page

117   "Processors were the rage": Bashe et al. 1986, p. 476.

120   "[Few] companies in U.S. business history": Sheehan 1956, p. 114.

121   "The least threatening of I.B.M.'s seven major rivals . . .": Burck 1964, p. 198.

121   "It doesn't do much good": Sobel 1981, p. 163.

122   "too smart not to let us": Burck 1964, p. 202.

124   "we are going to wind up with chaos": Quoted in Pugh et al. 1991, p. 119.

125   "orders not to come back": Watson 1990, p. 348.

125   "The problem was": Wise 1966a, p. 228.

126   "not even the Manhattan Project" and "was only half joking": Ibid., pp. 118–19.

126   "issued a memorandum suggesting": Rogers 1969, p. 272.

126   "Ordinary plants in those days": Watson 1990, p. 350.

127   "betokening all points of the compass": Pugh et al. 1991, p. 167.

128   "At the end of the risk assessment meeting": Wise 1966b, p. 205.

128   "six new computers and 44 new peripherals": Ibid., p. 138.

128   "the most important product announcement": Ibid., p. 138.

128   "The new System/360" and "most crucial": Wise 1966a, p. 119.

129   "one of this country's greatest industrial innovations": Quoted in DeLamarter 1986, p. 60.

130   "would become a standard code": Quoted in Campbell-Kelly 1989, p. 232.

130   "started to copy the machine from the skin in": Fishman 1981, p. 182.

133   "To have competitive leadership": Quoted in Pugh et al. 1991, p. 484.

133   "displace the 370 line": Quoted in ibid., p. 542.

133   "People involved with the project agree": Ibid., p. 550.

134   "In terms of sheer effort": Ibid., p. 549.

134   "the most expensive development-effort failure in IBM's history": Ibid., p. 550.

136   "We don't sell a product . . .": Burck 1964, p. 196.

CHAPTER 7. REAL TIME: REAPING THE WHIRLWIND

An excellent overview that sets Whirlwind, SAGE, and other military computer developments in the Cold War context is Paul Edwards's *The Closed World: Computers and the Politics of Discourse in Cold War America* (1996). Stuart W. Leslie's *The Cold War and American Science* (1993) also places these developments in a broad political and administrative framework. The detailed history of Project Whirlwind is the subject of Kent C. Redmond and Thomas M. Smith's monograph *Project Whirlwind: The History of a Pioneer Computer* (1980), while their subsequent book *From Whirlwind to MITRE: The R&D Story of the SAGE Air Defense Computer* (2000) completes the story. An excellent source on SAGE is a special issue of the *Annals of the History of Computing* (5, no. 4, 1981). John F. Jacobs's *The SAGE Air Defense System* (1986) is an interesting personal account (the book was published by the MITRE Corporation, of which Robert E. Everett became president).

There are several good accounts of American Airlines' SABRE. The most recent, and best, appears in James L. McKenney's *Waves of Change* (1995). There is no full-length history of the bar code, but we have made good use of Alan Q. Morton's excellent article "Packaging History: The Emergence of the Uniform Product Code" (1994). Another useful source is Steven Brown's personal account of the negotiations to implement bar codes in the U.S. grocery industry, *Revolution at the Checkout Counter* (1997).

Page

144 **"turned on a light"**: Redmond and Smith 1980, p. 33.

144 **Crawford gave a lecture:** Reprinted in Campbell-Kelly and Williams 1985, pp. 375–92.

144 **"gold-plated boondoggle"**: Quoted in Redmond and Smith 1980, p. 38.

144 **"Oh we were cocky!"**: Ibid.

144 **"solution of many scientific and engineering problems"**: Ibid., p. 41.

145 **"the tail had passed"**: Ibid., p. 46.

146 **"saw no reason to allocate time"**: Ibid., p. 47.

147 **"fearless and imaginative jumps"**: Ibid., p. 96.

147 **"Jay took a bunch of stuff"**: Quoted in ibid., p. 183.

148 **"fantastic," "appalling," and further quotations:** Quoted in ibid., pp. 145, 150.

148 **"lame, purblind, and idiot-like"**: Quoted in ibid., p. 172.

149 **"I remembered having heard" and "differed only in their degree of negativity"**: Valley 1985, pp. 207–8.

150 **49,000 tubes and weighing 250 tons:** Jacobs 1986, p. 74.

151 **"expensive white elephant"**: Flamm 1987, p. 176

151 **The real contribution of SAGE:** See Flamm 1987 and 1988.

151 **"the chances [were] reasonably high"**: Bennington 1983, p. 351.

151 **"It was the Cold War that helped IBM"**: Watson and Petre 1990, p. 230.

154 **Reservisor:** For an account of the Reservisor, see Eklund 1994.

154 **"most business travelers"**: McKenney 1995, p. 98.

155 **"tough applications, such as Banks"**: Quoted in Bashe et al. 1986, p. 5.

156 **"You'd better make those black boxes do the job"**: Quoted in McKenney 1995, p. 111.

156 **"85,000 daily telephone calls"**: Plugge and Perry 1961, p. 593.

157 **"the kids' SAGE"**: Burck 1965, p. 34.

174   "classic data-processing calamities": McKenney 1995, p. 119.
178   "a blue ribbon group": Quoted in Morton 1994, p. 105

CHAPTER 8. SOFTWARE

The literature of the history of software is patchy. Although good for its kind, much of it is "internalist," technical, and focused on narrow software genres, which makes it difficult for a nonexpert to get a foothold in the topic. An excellent starting point is Steve Lohr's engaging biographically oriented *Go To: The Story of the Programmers Who Created the Software Revolution* (2001). To try to bring some order to the history of the software field, a conference was organized in 2000 by the Heinz Nixdorf MuseumsForum, Paderborn, Germany, and Charles Babbage Institute, University of Minnesota. The papers presented at the conference were published as *History of Computing—Software Issues* (Hashagen et al., eds., 2002). This book gives the best overview of the subject so far published.

Saul Rosen's *Programming Systems and Languages* (1967) gives about the best coverage of the early period. There is a very full literature on the history of programming languages, of which the best books are Jean Sammet's *Programming Languages: History and Fundamentals* (1969) and the two edited volumes of the History of Programming Languages conferences that took place in 1976 and 1993: *History of Programming Languages* (Richard Wexelblat, ed., 1981), and *History of Programming Languages*, volume 2 (Thomas Bergin et al., eds., 1996). The extensive literature on the history of programming languages has distorted the view of the history of software, so that most other aspects are neglected by comparison. The OS/360 debacle is discussed at length in Fred Brooks's classic *The Mythical Man-Month* (1975) and in Emerson W. Pugh et al.'s *IBM's 360 and Early 370 Systems* (1991). There is no monographic history of the software crisis, but Peter Naur and Brian Randell's edited proceedings of the Garmisch conference, *Software Engineering* (1968), is rich in sociological detail.

The two best books on the history of the software industry are Martin Campbell-Kelly's *From Airline Reservations to Sonic the Hedgehog: A History of the Software Industry* (2003), which focuses on the U.S. scene, and David Mowery's edited volume, *The International Computer Software Industry* (1996).

Page
165   **Wilkes, began to turn his thoughts . . . toward the programming problem:** See Campbell-Kelly 1982.
166   **"By June 1949 people had begun to realize":** Wilkes 1985, p. 145.
168   *The Preparation of Programs for an Electronic Digital Computer:* Wilkes, Wheeler, and Gill 1951.

169   "programming and debugging": Backus 1979, p. 22.

170   "on the nineteenth floor . . .": Ibid., p. 24.

170   "[T]hey had heard too many glowing descriptions": Ibid., p. 27.

170   "often we would rent rooms": Ibid., p. 29.

170   "a euphemism for April 1957": Rosen 1967, p. 7.

171   "SOURCE PROGRAM ERROR . . ." and "Flying blind": Bright 1979, p. 73.

171   the first FORTRAN textbook: McCracken 1961.

172   "missionary zeal for the cause": Rosen 1967, p. 12.

173   Hundreds of languages: See Sammet 1969.

174   System Development Corporation: See Baum 1981.

175   "all one needed was a coding pad and a pencil": Quoted in Campbell-Kelly 1995, p. 83.

178   "the centerpiece of the programming support": Pugh et al. 1991, p. 326.

178   "Yesterday a key man was sick": Brooks 1975, p. 154.

179   "realistic," "painfully sluggish," and "was in trouble": Pugh et al. 1991, pp. 336, 340, 342.

179   "A few months ago IBM's software budget": Quoted in Watson 1990, p. 353.

179   "growing mood of desperation": Pugh et al. 1991, p. 336.

179   "Like dousing a fire with gasoline" and "The bearing of a child": Brooks 1975, pp. 14, 17.

180   "the single largest cost": Watson 1990, p. 353.

180   "The cost to IBM": Pugh et al. 1991, p. 344.

180   On the morning of 22 July 1962: The best account of the *Mariner I* disaster is in Ceruzzi 1989, pp. 202–7.

181   "The phrase 'software engineering'": Naur and Randell 1968, p. 13.

181   "they are the industrialists" and "We build systems like the Wright brothers": Ibid., p. 17.

183   According to IBM sources: Fisher et al. 1983, pp. 176–77.

183   "the major event": Quoted in Campbell-Kelly 1995, p. 89.

184   "astounded": Ibid., p. 92.

184   "legitimate products with strong sales": Ibid.

CHAPTER 9. NEW MODES OF COMPUTING

Judy E. O'Neill's Ph.D. dissertation, "The Evolution of Interactive Computing Through Time-Sharing and Networking" (1992), gives the best overview of the development of time-sharing. Time-sharing at MIT was the subject of a double special issue of *Annals of the History of Computing*, edited by J. A. N. Lee (1992a and 1992b). The best account of the Dartmouth Time-Sharing System is John Kemeny and Thomas Kurtz's article "Dartmouth Time-Sharing," published in *Science* (1968). The history of Dartmouth BASIC is described in Kemeny's contributed chapter to Richard Wexelblat's *History of Programming Languages* (1981). A good historical account of Unix is Peter Salus's *A Quarter Century of Unix* (1994). Glyn Moody's *Rebel Code* (2001) is the best book so far on the Linux phenomenon.

A good overview of the minicomputer industry is given in Paul Ceruzzi's *History of Modern Computing* (2003). The most authoritative history of the Digital Equipment Corporation is *DEC Is Dead, Long Live DEC* (Edgar Schein et al. 2003); also useful is Glenn Rifkin and George Harrar's *The Ultimate Entrepreneur: The Story of Ken Olsen and Digital Equipment Corporation* (1985).

Good histories of the rise of the microelectronics industry are Ernest Braun and Stuart MacDonald's *Revolution in Miniature* (1982), P. R. Morris's *A History of the World Semiconductor Industry* (1990), and Andrew Goldstein and William Aspray's edited volume *Facets: New Perspectives on the History of Semiconductors* (1997). Also worth reading is Michael Malone's *The Microprocessor: A Biography* (1995). The Silicon Valley phenomenon is explored in AnnaLee Saxenian's *Regional Advantage: Culture and Competition in Silicon Valley and Route 128* (1994) and Martin Kenney's edited volume *Understanding Silicon Valley* (2000).

There is no monographic study of the calculator industry, although An Wang's *Lessons* (1986) and Edwin Darby's *It All Adds Up* (1968) give useful insights into the decline of the U.S. industry. The most analytical account of the early U.S. video game industry is Ralph Watkins's Department of Commerce report *A Competitive Assessment of the U.S. Video Game Industry* (1984). Of the popular books on the video game industry, Steven Poole's *Trigger Happy* (2000) and Scott Cohen's *Zap! The Rise and Fall of Atari* (1984) offer good historical insights. Leonard Herman's *Phoenix: The Fall and Rise of Home Videogames* (1997), self-published, contains a wealth of chronologically arranged data.

Page

188   "simple enough to allow": Kemeny and Kurtz 1968, p. 225.
191   "That would leave, say, five years": Licklider 1960, p. 132.
191   "variously translated as": Fano and Corbato 1966, p. 77.
191   "The intent of the summer session": Lee 1992a, p. 46.
193   "hottest new talk of the trade": Main 1967, p. 187.
193   "amazed," "overwhelmingly," and "irrational": Quoted in O'Neill 1992, pp. 113–14.
193   "evolve into a problem solving service ...": Burck 1968, pp. 142–43.
193   "Grosch's Law": See Ralston and Reilly 1993, p. 586; Grosch 1989, pp. 180–182.
194   "[b]arring unforeseen obstacles": Greenberger 1964, p. 67.
194   "And while it may seem odd": Baran 1970, p. 83.

195   "illusion" and "computer myths": Gruenberger 1971, p. 40.
195   "team of hundreds . . .": Main 1967, p. 187.
196   "We didn't want to lose": Ritchie 1984a, p. 1578.
197   "a somewhat treacherous pun": Ibid., p. 1580.
197   "scrounged around, coming up with": Slater 1987, p. 278.
197   "A mouse with an IBM operating system": See, for example, Barron 1971, p. 1.
198   "Because they were starting afresh": Ritchie 1984b, p. 758.
198   "The genius of the UNIX system": Ibid.
200   "father of venture capital": Gompers 1994, p. 5.
200   "could not believe that in 1960": Quoted in Fisher et al. 1983, p. 273.
203   "Moore's Law": Quotations from Moore 1965.

CHAPTER 10. THE SHAPING OF THE PERSONAL COMPUTER

A good overview of the early development of the personal computer is Paul Freiberger and Michael Swaine's *Fire in the Valley* (1984 and 1999). Stan Veit's *History of the Personal Computer* (1993) is full of anecdotes and insights on the early development of the industry to be found nowhere else; Veit participated in the rise of the personal computer, first as a retailer and then as editor of *Computer Shopper*. The later period is covered by Robert X. Cringely's *Accidental Empires* (1992); although less scholarly than some other books, it is generally reliable and is written with a memorable panache.

There have been several books written about most of the major firms in the personal-computer industry. For Apple Computer we found the most useful to be Jim Carlton's *Apple: The Inside Story* (1997) and Michael Moritz's *The Little Kingdom: The Private Story of Apple Computer* (1984). There is a seemingly inexhaustible supply of histories of Microsoft, few of which have anything noteworthy to add to what is already well known. We have found the most useful on the early years to be: Stephen Manes and Paul Andrews's *Gates: How Microsoft's Mogul Reinvented an Industry* (1994), James Wallace and Jim Erickson's *Hard Drive: Bill Gates and the Making of the Microsoft Empire* (1992), and Daniel Ichbiah and Susan L. Knepper's *The Making of Microsoft* (1991). The best account of the creation of the IBM PC is James Chposky and Ted Leonsis's *Blue Magic* (1988). The following books contain much useful information about the industry as well as individuals: Robert Levering et al.'s *The Computer Entrepreneurs* (1984) and Robert Slater's *Portraits in Silicon* (1987).

In pursuing the analogy between the development of radio broadcasting and the personal computer, we have been guided by Susan J. Douglas's *Inventing American Broadcasting* (1987), Susan Smulyan's *Selling Radio* (1994), and Erik Barnouw's *A Tower in Babel* (1966).

Page

210   "Announcing a new era of integrated electronics": The ad is reproduced in Augarten 1984, p. 264.

211   "fad" that "seemed to come from nowhere": Douglas 1987, p. 303.

213–214 "Exclusive! Altair 8800.": *Popular Electronics,* January 1975, p. 33; reproduced in Langlois 1992, p. 10.

219   "nit-picking technical debates": Moritz 1984, p. 136.

220   three leading manufacturers: See Langlois 1992 for an excellent economic analysis of the personal-computer industry.

221   "The home computer that's ready to work, play and grow with you . . . .": Quoted in Moritz 1984, p. 224.

222   There were three main markets for applications software: for a discussion of the personal-computer software industry, see Campbell-Kelly 2003, chapter 7.

222   "When customers walked into computer stores in 1979": Freiberger and Swaine 1984, p. 135.

224   "Suddenly it became obvious": Slater 1987, pp. 285–86.

227   "Is he Mary Gates's son?": Ichbiah and Knepper 1991, p. 77.

228   "Why on earth would you care about the personal computer?": Chposky and Leonsis 1988, p. 107.

228   "cold and aloof": Ibid., p. 80.

228   "a human face": *Time,* 11 July 1983, quoted in Chposky and Leonsis 1988, p. 80.

CHAPTER 11. BROADENING THE APPEAL

The fullest account of the history of the personal-computer software industry appears in Martin Campbell-Kelly's *From Airline Reservations to Sonic the Hedgehog* (2003). Although the numerous histories of Microsoft (of which several were listed in the notes to chapter 10) take a Microsoft-centric view of the software universe, they also consider some of its competitors.

The history of the graphical user interface is authoritatively covered in *A History of Personal Workstations,* edited by Adele Goldberg (1988). The inside story of Xerox PARC is told in Michael Hiltzik's *Dealers of Lightning: Xerox PARC and the Dawn of the Computer Age* (1999) and in Douglas Smith and Robert Alexander's earlier *Fumbling the Future: How Xerox Invented, Then Ignored, the First Personal Computer* (1988). Thierry Bardini's *Bootstrapping: Douglas Engelbart, Coevolution, and the Origins of Personal Computing* (2000) at last does justice to a once-unsung hero of the personal-computer revolution.

Books focusing on the Macintosh development include John Sculley's insider account *Odyssey: Pepsi to Apple* (1987) and Steven Levy's external view *Insanely Great*

(1994). A selection of business histories of Microsoft (all of which discuss the Windows operating system) was listed in the notes to chapter 10. We found the most thoughtful account of Microsoft's forays into CD-ROM publishing and consumer networks to be Randall Stross's *The Microsoft Way* (1996).

The history of consumer networks has largely been sidelined by the rise of the Internet. One exception is Kara Swisher's *Aol.com* (1999), which examines AOL's earlier history as well as its meteoric growth in the 1990s. There are no substantial histories of early players CompuServe, The Source, Prodigy, or Genie, although they appear fleetingly in Swisher's *Aol.com*.

Page

233 "cut it off and throw you in the pool": Quoted in Campbell-Kelly 1995, p. 103. the total market for personal-computer software (and other statistics): Ibid., p. 98.

233 "marketing, marketing, and marketing": Quoted in Sigel 1984, p. 126.

234 The story of the Lotus Development Corporation: See Petre 1985.

237 "[t]he mouse consistently beat out the other devices" and "[n]one of us would have thought": Quoted in Goldberg 1988, pp. 195–96.

238–39 "it made a profound impression," "two threads," "architecture of information," "the preferred style," and "it was hard for people to believe": Lampson 1988, pp. 294–95.

240 "Why isn't Xerox marketing this? . . .": Smith and Alexander 1988, p. 241.

240 the idea of Jef Raskin: See interview with Raskin in Lammers 1986, pp. 227–45.

241 "Steve's 'pirates'": Sculley 1987, p. 157.

241 "Apple Computer was about to introduce its Macintosh computer": Wallace and Erickson 1992, p. 267.

242 "People weren't about to buy $2,000 computers": Sculley 1987, p. 248.

243 several firms had already started: See Markoff 1984.

244 "basically died on the vine": Slater 1987, p. 260.

244 dubbed it "TopHeavy" . . . ": Carroll 1994, p. 86.

244 "to have our name basically define the generic": Quoted in Wallace and Erickson 1992, p. 252.

245 "unbearably slow": Ichbiah and Knepper 1991, p. 189.

245 "the Company's registered audiovisual copyrights": This form of words appears in Apple Computer's annual reports.

246 "[s]ome 6,000 people" and "the most extravagant": Ichbiah and Knepper 1991, p. 239.

247 "Everyone knows encyclopedias": Lammers 1986, p. 69.

249 "How long does it take": Quoted in Campbell-Kelly 2003, p. 294.

250 "A CompuServe subscriber": Levering et al. 1984, pp. 415–6.

251   "an easy-to-use point-and-click interface anyone's grandmother could install": Swisher 1998, p. 66.
251   "carpet bombing": Swisher 1998, p. 99.
252   "In 1993, the media had carried stories": Stross 1996, p. 168.

CHAPTER 12. FROM THE WORLD BRAIN TO THE WORLD WIDE WEB

Since the first edition of this book appeared, there has been a flurry of general histories of the Internet, several of which are excellent. We recommend Janet Abbate's *Inventing the Internet* (1999), John Naughton's *A Brief History of the Future* (2001), Michael and Ronda Hauben's *Netizens: On the History and Impact of Usenet and the Internet* (1997), and Christos Moschovitis's *History of the Internet* (1999). For Internet statistics we have used data published by the Internet Software Consortium (www.isc.org).

The historical context of the World Brain is given in a new edition of Wells's 1938 classic, edited by Alan Mayne: *World Brain: H. G. Wells on the Future of World Education* (1995). Historical accounts of Bush's memex are given in James M. Nyce and Paul Kahn's edited volume *From Memex to Hypertext: Vannevar Bush and the Mind's Machine* (1991) and Colin Burke's *Information and Secrecy: Vannevar Bush, Ultra, and the Other Memex* (1994).

The history of the DARPA Information Processing Techniques Office (IPTO), which effectively created the Internet, is very fully described in Arthur L. Norberg and Judy E. O'Neill's *Transforming Computer Technology: Information Processing for the Pentagon, 1962–1986* (2000) and Alex Roland and Philip Shiman's *Strategic Computing: DARPA and the Quest for Machine Intelligence, 1983–1993* (2002). Mitchell Waldrop has written a fine biography of J. C. R. Licklider, IPTO's founder and the seminal figure of personal computing: *The Dream Machine: J.C.R. Licklider and the Revolution That Made Computing Personal* (2001).

The context and evolution of the World Wide Web has been described by its inventor Tim Berners-Lee in *Weaving the Web* (1999), and by his colleagues at CERN James Gillies and Robert Cailliau in *How the Web Was Born* (2000). The "browser wars" are recounted in detail in Michael Cusumano and David Yoffie's *Competing on Internet Time* (1998).

Histories of the commercial development of the World Wide Web are beginning to appear. The early days are chronicled in Robert Reid's *Architects of the Web: 1,000 Days That Built the Future of Business* (1997), while the crash is recounted in John Cassidy's *Dot.con* (2003). As individual enterprises rise to prominence, business histories of them soon appear. We found the following useful: Robert Spector's *Amazon.com: Get Big Fast* (2000), Karen Angel's *Inside Yahoo!* (2001), and Joseph Menn's *All the Rave: The Rise and Fall of Shawn Fanning's Napster* (2003).

One day Internet regulation will be a topic of major historical importance. At the time of writing, however, history is unfolding and very little history has therefore been written. Of the contemporary discussions and debates we have found Lawrence Lessig's *Code and Other Laws of Cyberspace* (1999) and Jessica Litmans's *Digital Copyright: Protecting Intellectual Property on the Internet* (2001) to be among the most thought provoking.

Page

256   "pushed aside by men like Hitler": Wells 1938, p. 46.

257   "A World Encyclopedia no longer presents itself": Ibid., pp. 48–49.

257   "great number of workers": Ibid., p. 60.

257   Wells was invited to lunch: See Smith 1986.

257   "in a couple of cubic feet": Bush 1945.

258   "a library of a million volumes" and following quotations: Ibid.

259   "in the future, but not so far": Bush 1967, p. 99.

259   "however unworthy it may be, to Dr. Bush": Licklider 1965, p. xiii.

259   "found this article in *Life* magazine": Quoted in Goldberg 1988, pp. 235–36.

260   "It seems reasonable to envision": Licklider 1960, p. 135.

261   "his enthusiasm infected me": Goldberg 1988, p. 144.

261   the idea was first put forward: See Campbell-Kelly 1988.

262   "Suddenly I learned how to route packets": Quoted in Abbate 1994, p. 41.

264   "There was more than one person": Quoted in ibid., p. 1.

264   "When the mail [program] was being developed": Quoted in ibid., p. 82.

265   "the Great Electronic Mail Shootout": A headline in *Fortune*, 20 August 1984.

269   "the marriage of hypertext and the Internet": Berners-Lee 1999, p. 28.

270   "every project on display would have something to do with the Web": Ibid., p. 56.

273   "embrace and extend" and "was willing to sacrifice one child": Cusumano and Yoffie 1998, pp. 08 and 112.

275   "The time is close at hand": Wells 1938, p. 54.

277   "he left his day job": Cassidy 2003, p. 163.

■■ BIBLIOGRAPHY

Abbate, Janet. 1994. "From Arpanet to Internet: A History of ARPA-Sponsored Computer Networks, 1966–1988." Ph.D. diss., University of Pennsylvania.

————. 1999. *Inventing the Internet*. Cambridge, Mass.: MIT Press.

Agar, Jon. 2003. *The Government Machine: A Revolutionary History of the Computer*. Cambridge, Mass.: MIT Press.

Aiken, Howard H. 1937. "Proposed Automatic Calculating Machine." In Randell 1982, pp. 195–201.

Aiken, Howard H., et al. 1946. *A Manual of Operation of the Automatic Sequence Controlled Calculator*. Cambridge, Mass.: Harvard University Press. Reprint, with a foreword by I. Bernard Cohen and an introduction by Paul Ceruzzi, volume 8, Charles Babbage Institute Reprint Series for the History of Computing, Cambridge, Mass., and Los Angeles: MIT Press and Tomash Publishers, 1985.

Alborn, Tim. 1994. "Public Science, Private Finance: The Uneasy Advancement of J. W. Lubbock." In *Proceedings of a Conference on Science and British Culture in the 1830s* (pp. 5–14), 6–8 July, Trinity College, Cambridge.

Allyn, Stanley C. 1967. *My Half Century with NCR*. New York: McGraw-Hill.

Anderson, Margo J. 1988. *The American Census: A Social History*. New Haven, Conn.: Yale University Press.

Angel, Karen. 2001. *Inside Yahoo! Reinvention and the Road Ahead*. New York: John Wiley.

Anonymous. 1874. "The Central Telegraph Office." *Illustrated London News*, 28 November, p. 506, and 10 December, p. 530.

————. 1920. "The Central Telegraph Office, London." BT Archives, London, POST 82/66.

————. 1932. "International Business Machines." *Fortune*, January, pp. 34–50.

————. 1940. "International Business Machines." *Fortune*, January, p. 36.

————. 1950. "Never Stumped: International Business Machines' Selective Sequence Electronic Calculator." *The New Yorker*, 4 March, pp. 20–21.

————. 1952. "Office Robots." *Fortune*, January, p. 82.

Ashford, Oliver M. 1985. *Prophet—or Professor? The Life and Work of Lewis Fry Richardson.* Boston: Adam Hilger.

Aspray, William. 1990. *John von Neumann and the Origins of Modern Computing.* Cambridge, Mass.: MIT Press.

Aspray, William, ed. 1990. *Computing Before Computers.* Ames: Iowa State University Press.

Augarten, Stan. 1984. *Bit by Bit: An Illustrated History of Computers.* New York: Ticknor & Fields.

Austrian, Geoffrey D. 1982. *Herman Hollerith: Forgotten Giant of Information Processing.* New York: Columbia University Press.

Babbage, Charles. 1989. *Works of Babbage.* Ed. M. Campbell-Kelly. 11 vols. New York: American University Press.

————. 1822a. "A Letter to Sir Humphrey Davy." In Babbage 1989, vol. 2, pp. 6–14.

————. 1822b. "The Science of Number Reduced to Mechanism." In Babbage 1989, vol. 2, pp. 15–32.

————. 1834. "Statement Addressed to the Duke of Wellington." In Babbage 1989, vol. 3, pp. 2–8.

————. 1835. *Economy of Machinery and Manufactures.* In Babbage 1989, vol. 5.

————. 1852. "Thoughts on the Principles of Taxation." In Babbage 1989, vol. 5, pp. 31–56.

————. 1994. *Passages from the Life of a Philosopher.* Edited with a new introduction by Martin Campbell-Kelly. 1864. New Brunswick, N.J.: IEEE Press and Rutgers University Press. Also in Babbage 1989, vol. 11.

Backus, John. 1979. "The History of FORTRAN I, II, and III." *Annals of the History of Computing* 1, no. 1: 21–37.

Baran, Paul. 1970. "The Future Computer Utility." In Taviss 1970, pp. 81–92.

Bardini, Thierry. 2000. *Bootstrapping: Douglas Engelbart, Coevolution, and the Origins of Personal Computing.* Stanford, Calif.: Stanford University Press.

Barnett, C. C., Jr., B. R. Anderson, W. N. Bancroft et al. 1967. *The Future of the Computer Utility.* New York: American Management Association.

Barnouw, Erik. 1967. *A Tower in Babel (History of Broadcasting in the United States).* New York: Oxford University Press.

Barron, David W. 1971. *Computer Operating Systems.* London: Chapman and Hall.

Bashe, Charles J., Lyle R. Johnson, John H. Palmer, and Emerson W. Pugh. 1986. *IBM's Early Computers.* Cambridge, Mass.: MIT Press.

Baum, Claude. 1981. *The System Builders: The Story of SDC.* Santa Monica, Calif.: System Development Corporation.

Baxter, James Phinney. 1946. *Scientists Against Time.* Boston: Little, Brown.

Beeching, Wilfred A. 1974. *A Century of the Typewriter.* London: Heinemann.

Belden, Thomas G., and Marva R. Belden. 1962. *The Lengthening Shadow: The Life of Thomas J. Watson.* Boston: Little, Brown.

Beniger, James R. 1986. *The Control Revolution: Technological and Economic Origins of the Information Society.* Cambridge, Mass.: Harvard University Press.

Bennington, Herbert D. 1983. "Production of Large Computer Programs" (with new introduction). 1956. *Annals of the History of Computing* 5, no. 4: 350–61.

Bergin, Thomas J., Richard G. Gibson, and Richard G. Gibson Jr., eds. 1996. *History of Programming Languages,* vol. 2. Reading, Mass.: Addison-Wesley.

Berkeley, Edmund Callis. 1949. *Giant Brains or Machines That Think.* New York: John Wiley.

Berners-Lee, Tim. 1999. *Weaving the Web: The Past, Present, and Future of the World Wide Web by Its Inventor.* London: Orion Business Books.

Bernstein, Jeremy. 1981. *The Analytical Engine.* New York: Random House.

Bliven, Bruce, Jr., 1954. *The Wonderful Writing Machine.* New York: Random House.

Bowden, B. V., ed. 1953. *Faster Than Thought.* London: Pitman.

Braun, Ernest, and Stuart Macdonald. 1978. *Revolution in Miniature: The History and Impact of Semiconductor Electronics.* Cambridge: Cambridge University Press.

Brennan, Jean F. 1971. *The IBM Watson Laboratory at Columbia University: A History.* New York: IBM Corp.

Bright, Herb. 1979. "FORTRAN Comes to Westinghouse-Bettis, 1957." *Annals of the History of Computing* (July): 72–74.

Brock, Gerald W. 1975. *The U.S. Computer Industry: A Study of Market Power.* Cambridge, Mass.: Ballinger.

Bromley, Alan G. 1982. "Charles Babbage's Analytical Engine, 1838." *Annals of the History of Computing* 4, no. 3: 196–217.

———. 1990a. "Difference and Analytical Engines." In Aspray 1990, 59–98.

———. 1990b. "Analog Computing Devices." In Aspray 1990, 156–99.

Brooks, Frederick P., Jr. 1975. *The Mythical Man-Month: Essays in Software Engineering.* Reading, Mass.: Addison-Wesley.

Brown, Steven A. 1997. *Revolution at the Checkout Counter: The Explosion of the Bar Code.* Cambridge, Mass.: Harvard University Press.

Bud-Frierman, Lisa, ed. 1994. *Information Acumen: The Understanding and Use of Knowledge in Modern Business.* London and New York: Routledge.

Burck, Gilbert. 1964. "The Assault on Fortress I.B.M." *Fortune,* June, p. 112.

———. 1965. *The Computer Age and Its Potential for Management.* New York: Harper Torchbooks.

———. 1968. "The Computer Industry's Great Expectations." *Fortune,* August, p. 92.

Burke, Colin B. 1994. *Information and Secrecy: Vannevar Bush, Ultra, and the Other Memex.* Metuchen, N.J.: Scarecrow Press.

Burks, Alice R., and Arthur W. Burks. 1988. *The First Electronic Computer: The Atanasoff Story.* Ann Arbor: University of Michigan Press.

Bush, Vannevar. 1945. "As We May Think." *Atlantic Monthly,* July, pp. 101–8. Reprinted in Goldberg 1988, pp. 237–47.

————. 1967. *Science Is Not Enough*. New York: Morrow.

————. 1970. *Pieces of the Action*. New York: Morrow.

Campbell-Kelly, Martin. 1982. "Foundations of Computer Programming in Britain 1945–1955." *Annals of the History of Computing* 4: 133–62.

————. 1988. "Data Communications at the National Physical Laboratory (1965–1975)." *Annals of the History of Computing* 9, no. 3–4: 221–47.

————. 1989. *ICL: A Business and Technical History*. Oxford: Oxford University Press.

————. 1992. "Large-Scale Data Processing in the Prudential, 1850–1930." *Accounting, Business and Financial History* 2, no. 2: 117–39.

————. 1994. "The Railway Clearing House and Victorian Data Processing." In Bud-Frierman 1994, pp. 51–74.

————. 1995. "The Development and Structure of the International Software Industry, 1950–1990." *Business and Economic History* 24, no. 2: 73–110.

————. 1996. "Information Technology and Organizational Change in the British Census, 1801–1911." *Information Systems Research* 7, no. 1: 22–36. Reprinted in Yates and Van Maanen 2001, pp. 35–58.

————. 1998. "Data Processing and Technological Change: The Post Office Savings Bank, 1861–1930." *Technology and Culture* 39: 1–32.

————. 2003. *From Airline Reservations to Sonic the Hedgehog: A History of the Software Industry*. Cambridge, Mass.: MIT Press.

Campbell-Kelly, Martin, Mary Croarken, Eleanor Robson, and Raymond Flood, eds. 2003. *Sumer to Spreadsheets: The History of Mathematical Tables*. Oxford: Oxford University Press.

Campbell-Kelly, Martin, and Michael R. Williams, eds. 1985. *The Moore School Lectures*. Cambridge, Mass., and Los Angeles: MIT Press and Tomash Publishers.

Cannon, James G. 1910. *Clearing Houses*. Washington, D.C.: National Monetary Commission.

Carlton, Jim. 1997. *Apple: The Inside Story of Intrigue, Egomania, and Business Blunders*. New York: Random House.

Carroll, Paul. 1994. *Big Blues: The Unmaking of IBM*. London: Weidenfeld & Nicolson.

Cassidy, John. 2003. *Dot.con: The Real Story of Why the Internet Bubble Burst*. London: Penguin.

Ceruzzi, Paul E. 1983. *Reckoners: The Prehistory of the Digital Computer, from Relays to the Stored Program Concept, 1935–1945*. Westport, Conn.: Greenwood Press.

————. 1989. *Beyond the Limits: Flight Enters the Computer Age*. Cambridge, Mass.: MIT Press.

————. 1991. "When Computers Were Human." *Annals of the History of Computing* 13, no. 3: 237–44.

————. 1998. *A History of Modern Computing*. Cambridge, Mass.: MIT Press.

Chase, G. C. 1980. "History of Mechanical Computing Machinery." *Annals of the History of Computing* 2, no. 3: 198–226. Reprint of 1952 conference paper, with an introduction by I. B. Cohen.

Chposky, James, and Ted Leonsis. 1988. *Blue Magic: The People, Power and Politics Behind the IBM Personal Computer.* New York: Facts on File.

Cohen, I. Bernard. 1988. "Babbage and Aiken." *Annals of the History of Computing* 10, no. 3: 171–93.

———. 1999. *Howard Aiken: Portrait of a Computer Pioneer.* Cambridge, Mass.: MIT Press.

Cohen, Patricia Cline. 1982. *A Calculating People: The Spread of Numeracy in Early America.* Chicago: University of Chicago Press.

Cohen, Scott. 1984. *Zap! The Rise and Fall of Atari.* New York: McGraw-Hill.

Collier, Bruce, and James MacLachlan. 1998. *Charles Babbage and the Engines of Perfection.* New York: Oxford University Press.

Comrie, L. J. 1933. "Computing the 'Nautical Almanac.'" *Nautical Magazine,* July, pp. 1–16.

———. 1946. "Babbage's Dream Comes True." *Nature* 158: 567–68.

Cortada, James W. 1993. *Before the Computer: IBM, NCR, Burroughs, and Remington Rand and the Industry They Created, 1865–1956.* Princeton, N.J.: Princeton University Press.

———. 1993. *The Computer in the United States: From Laboratory to Market, 1930–1960.* Armonk, N.Y.: M. E. Sharpe.

Cringley, Robert X. 1992. *Accidental Empires: How the Boys of Silicon Valley Make Their Millions, Battle Foreign Competition, and Still Can't Get a Date.* Reading, Mass.: Addison-Wesley.

Croarken, M. 1989. *Early Scientific Computing in Britain.* Oxford: Oxford University Press.

———. 1991. "Case 5656: L. J. Comrie and the Origins of the Scientific Computing Service Ltd." *IEEE Annals of the History of Computing* 21, no. 4: 70–71.

———. 2000. "L. J. Comrie: A Forgotten Figure in the History of Numerical Computation." *Mathematics Today* 36, no. 4 (August): 114–18.

Crowther, Samuel. 1923. *John H. Patterson: Pioneer in Industrial Welfare.* New York: Doubleday, Page.

Cusumano, Michael A., and Richard W. Selby. 1995. *Microsoft Secrets: How the World's Most Powerful Software Company Creates Technology, Shapes Markets, and Manages People.* New York: Free Press.

Cusumano, Michael A., and David B. Yoffie. 1998. *Competing on Internet Time: Lessons from Netscape and Its Battle with Microsoft.* New York: Free Press.

d'Ocagne, Maurice. 1986. *Le Calcul Simplifié.* Trans. J. Howlett and M. R. Williams. 1928. Cambridge, Mass.: MIT Press.

Darby, Edwin. 1968. *It All Adds Up: The Growth of Victor Comptometer Corporation.* Chicago: Victor Comptometer Corp.

David, Paul A. 1986. "Understanding the Economics of QWERTY: The Necessity of History." In Parker 1986, pp. 30–49.

Davies, Margery W. 1982. *Woman's Place Is at the Typewriter: Office Work and Office Workers, 1870–1930.* Philadelphia: Temple University Press.

DeLamarter, Richard Thomas. 1986. *Big Blue: IBM's Use and Abuse of Power.* New York: Dodd, Mead.

Douglas, Susan J. 1987. *Inventing American Broadcasting, 1899–1922.* Baltimore, Md.: Johns Hopkins University Press.

Eames, Charles, and Ray Eames. 1973. *A Computer Perspective.* Cambridge, Mass.: Harvard University Press.

Eckert, J. Presper. 1976. "Thoughts on the History of Computing." *Computer*, December, pp. 58–65.

Edwards, Paul N. 1996. *The Closed World: Computers and the Politics of Discourse in Cold War America.* Cambridge, Mass.: MIT Press.

Eklund, Jon. 1994. "The Reservisor Automated Airline Reservation System: Combining Communications and Computing." *Annals of the History of Computing* 16, no. 1: 6–69.

Engelbart, Doug. 1988. "The Augmented Knowledge Workshop." In Goldberg 1988, pp. 187–232.

Engelbart, Douglas C., and William K. English. 1968. "A Research Center for Augmenting Human Intellect." *Proceedings of the AFIPS 1968 Fall Joint Computer Conference* (pp. 395–410). Washington D.C.: Spartan Books.

Engelbourg, Saul. 1976. *International Business Machines: A Business History.* New York: Arno Press.

Engler, George Nichols. 1969. *The Typewriter Industry: The Impact of a Significant Technological Revolution.* Los Angeles: University of California, Los Angeles. Available from University Microfilms International, Ann Arbor, Mich.

Fano, R. M., and P. J. Corbato. 1966. "Time-Sharing on Computers." *Scientific American* pp. 76–95.

Fisher, Franklin M., James N.V. McKie, and Richard B. Mancke. 1983. *IBM and the U.S. Data Processing Industry: An Economic History.* New York: Praeger.

Fishman, Katherine Davis. 1981. *The Computer Establishment.* New York: Harper & Row.

Flamm, Kenneth. 1987. *Targeting the Computer: Government Support and International Competition.* Washington, D.C.: Brookings Institution.

———. 1988. *Creating the Computer: Government, Industry, and High Technology.* Washington, D.C.: Brookings Institution.

Foreman, R. 1985. *Fulfilling the Computer's Promise: The History of Informatics, 1962–1968.* Woodland Hills, Calif.: Informatics General Corp.

Forrester, Jay. 1971. *World Dynamics.* Cambridge, Mass.: Wright-Allen Press.

Frieberger, Paul, and Michael Swaine. 1999. *Fire in the Valley: The Making of the Personal Computer,* 2nd ed. New York: McGraw-Hill.

Gillies, James, and Robert Cailliau. 2000. *How the Web Was Born: The Story of the World Wide Web.* Oxford: Oxford University Press.

Goldberg, Adele, ed. 1988. *A History of Personal Workstations.* New York:

Goldstein, Andrew, and William Aspray, eds. 1997. *Facets: New Perspectives on the History of Semiconductors.* New York: IEEE Press.

Goldstine, Herman H. 1972. *The Computer: From Pascal to von Neumann.* Princeton, N.J.: Princeton University Press.

Grattan-Guinness, Ivor. 1990. "Work for the Hairdressers: The Production of de Prony's Logarithmic and Trigonometric Tables." *Annals of the History of Computing* 12, no. 3: 177–85.

Greenberger, Martin. 1964. "The Computers of Tomorrow." *Atlantic Monthly,* July, pp. 63–67.

Grier, David A. 1998. "The Math Tables Project: The Reluctant Start of the Computing Era." *IEEE Annals of the History of Computing* 20, no. 3: 33–50.

Grosch, Herbert R. J. 1989. *Computer: Bit Slices from a Life.* Lancaster, Pa.: Third Millennium Books.

Gruenberger, Fred, ed. 1971. *Expanding Use of Computers in the O's: Markets-Needs-Technology.* Englewood Cliffs, N.J.: Prentice-Hall.

Hashagen, Ulf, Reinhard Keil-Slawik, and Arthur Norberg, eds. 2002. *History of Computing-Software Issues.* Berlin, Germany: Springer-Verlag.

Hauben, Michael, and Ronda Hauben. 1997. *Netizens: On the History and Impact of Usenet and the Internet.* New York: Wiley-IEEE Computer Society Press.

Herman, Leonard. 1994. *Phoenix: The Fall and Rise of Home Videogames.* Union, N.J.: Rolenta Press.

Hessen, Robert. 1973. "Rand, James Henry, 1859–1944." *Dictionary of American Biography,* supplement 3, pp. 618–19.

Hiltzik, Michael. 1999. *Dealers of Lightning: Xerox PARC and the Dawn of the Computer Age.* New York: HarperBusiness.

Hoke, Donald R. 1990. *Ingenious Yankees: The Rise of the American System of Manufactures in the Private Sector.* New York: Columbia University Press.

Holcombe, Lee. 1973. *Victorian Ladies at Work: Middle-Class Working Women in England and Wales, 1850–1914.* Newton Abbot: David & Charles.

Hopper, Grace Murray. 1981. "The First Bug." *Annals of the History of Computing* 3, no. 3: 285–86.

Hyman, Anthony. 1982. *Charles Babbage: Pioneer of the Computer.* Princeton, N.J.: Princeton University Press.

Ichbiah, Daniel, and Susan L. Knepper. 1991. *The Making of Microsoft.* Rocklin, Calif.: Prima Publishing.

Jacobs, John F. 1986. *The SAGE Air Defense System: A Personal History.* Bedford, Mass.: MITRE Corp.

Kemeny, John G., and Thomas E. Kurtz. 1968. "Dartmouth Time-Sharing." *Science* 162, no. 3850 (11 October): 223–28.

Kenney, Martin, ed. 2000. *Understanding Silicon Valley: The Anatomy of an Entrepreneurial Region.* Palo Alto, Calif.: Stanford University Press.

Kieve, Jeffrey L. 1973. *Electric Telegraph: A Social and Economic History.* Newton Abbot: David & Charles.

Lammers, Susan. 1986. *Programmers at Work: Interviews with Nineteen Programmers Who Shaped the Computer Industry.* Washington, D.C.: Tempus, Redmond.

Lampson, Butler W. 1988. "Personal Distributed Computing." In Goldberg 1988, pp. 293–335.

Langlois, Richard N. 1992. "External Economies and Economic Progress: The Case of the Microcomputer Industry." *Business History Review* 66, no. 1: 1–50.

Lardner, Dionysius. 1834. "Babbage's Calculating Engine." In Babbage 1989, vol. 2, pp. 118–86.

Lavington, Simon H. 1975. *A History of Manchester Computers.* Manchester, England: NCC.

Lee, J. A. N., ed. 1992a and 1992b. "Special Issue: Time-Sharing and Interactive Computing at MIT." *Annals of the History of Computing* 14, nos. 1 and 2.

Leslie, Stuart W. 1983. *Boss Kettering.* New York: Columbia University Press.

———. 1993. *The Cold War and American Science: The Military-Industrial-Academic Complex at MIT and Stanford.* New York: Columbia University Press.

Lessig, Lawrence. 1999. *Code and Other Laws of Cyberspace.* New York: Basic Books.

Levering, Robert, Michael Katz, and Milton Moskowitz. 1984. *The Computer Entrepreneurs.* New York: New American Library.

Levy, Steven. 1994. *Insanely Great: The Life and Times of Macintosh, the Computer That Changed Everything.* New York: Viking.

Licklider, J. C. R. 1960. "Man-Computer Symbiosis." *IRE Transactions on Human Factors in Electronics* (March): 4–11. Also in Goldberg 1988, pp. 131–40.

———. 1965. *Libraries of the Future.* Cambridge, Mass.: MIT Press.

Lindgren, M. 1990. *Glory and Failure: The Difference Engines of Johann Muller, Charles Babbage, and Georg and Edvard Scheutz.* Cambridge, Mass.: MIT Press.

Litman, Jessica. 2001. *Digital Copyright: Protecting Intellectual Property on the Internet.* New York: Prometheus Books.

Lohr, Steve. 2001. *Go To: The Story of the Programmers Who Created the Software Revolution.* New York: Basic Books.

Longacre, W. F. 1906. "Systematizing a Large Business." *Business Man's Magazine and Book-Keeper,* p. 172.

Lovelace, Ada A. 1843. "Sketch of the Analytical Engine." In Babbage 1989, vol. 3, pp. 89–170.

Lukoff, Herman. 1979. *From Dits to Bits.* Portland, Ore.: Robotics Press.

Main, Jeremy. 1967. "Computer Time-Sharing—Everyman at the Console." *Fortune,* August, p. 88.

Malone, Michael S. 1995. *The Microprocessor: A Biography.* New York: Springer-Verlag.

Manes, Stephen, and Paul Andrews. 1994. *Gates: How Microsoft's Mogul Reinvented an Industry—and Made Himself the Richest Man in America.* New York: Simon & Schuster.

Marcosson, Isaac F. 1945. *Wherever Men Trade: The Romance of the Cash Register.* New York: Dodd, Mead.

Markoff, John. 1984. "Five Window Managers for the IBM PC." *Byte(/),* Fall: 65–87.

Martin, Thomas C. 1891. "Counting a Nation by Electricity." *Electrical Engineer* 12: 521–30.

Mauchly, John. 1942. "The Use of High Speed Vacuum Tube Devices for Calculating." In Randell 1982, pp. 355–58.

May, Earl C., and Will Oursler. 1950. *The Prudential: A Story of Human Security.* New York: Doubleday.

McCartney, Scott. 1999. *ENIAC: The Triumphs and Tragedies of the World's First Computer.* New York: Walker.

McCracken, Daniel D. 1961. *A Guide to FORTRAN Programming.* New York: John Wiley.

McKenney, James L., with Duncan G. Copeland and Richard Mason. 1995. *Waves of Change: Business Evolution Through Information Technology.* Boston: Harvard Business School Press.

Menn, Joseph. 2003. *All the Rave: The Rise and Fall of Shawn Fanning's Napster.* New York: Crown Business.

Metropolis, Nicholas, J. Howlett, and Gian-Carlo Rota, eds. 1980. *A History of Computing in the Twentieth Century.* New York: Academic Press.

Mollenhoff, Clark R. 1988. *Atanasoff: Forgotten Father of the Computer.* Ames: Iowa State University Press.

Moody, Glyn. 2001. *Rebel Code: Linux and the Open Source Revolution.* New York: Perseus.

Moore, Gordon E. 1965. "Cramming More Components onto Integrated Circuits." *Electronics* 38 (19 April): 114–17.

Moritz, Michael. 1984. *The Little Kingdom: The Private Story of Apple Computer.* New York: Morrow.

Morris, P. R. 1990. *A History of the World Semiconductor Industry.* London: Peter Perigrinus/IEE.

Morton, Alan Q. 1994. "Packaging History: The Emergence of the Uniform Product Code (UPC) in the United States, 1970–75." *History and Technology* 11: 101–11.

Moschovitis, Christos J. P., et al. 1999. *History of the Internet: A Chronology, 1843 to Present.* Santa Barbara, Calif.: ABC-CLIO.

Mowery, David C., ed. 1996. *The International Computer Software Industry.* New York: Oxford University Press.

Naughton, John. 2001. *A Brief History of the Future: From Radio Days to Internet Years in a Lifetime.* Woodstock and New York: Overlook.

Naur, Peter, and Brian Randell, eds. 1969. "Software Engineering." Report on a conference sponsored by the NATO Science Committee, Garmisch-Partenkirchen, Germany, 7–11 October 1968. Brussels: NATO Scientific Affairs Division.

NCR. 1984a. *NCR: 1884–1922: The Cash Register Era.* Dayton, Ohio: NCR.

———. 1984b. *NCR: 1923–1951: The Accounting Machine Era.* Dayton, Ohio: NCR.

———. 1984c. *NCR: 1952–1984: The Computer Age.* Dayton, Ohio: NCR.

———. 1984d. *NCR: 1985 and Beyond: The Information Society.* Dayton, Ohio: NCR.

Nebeker, Frederik. 1995. *Calculating the Weather: Meteorology in the Twentieth Century.* New York: Academic Press.

Nelson, Theodor H. 1974. *Computer Lib: You Can and Must Understand Computers Now.* Chicago: Theodor H. Nelson.

———. 1974. *Dream Machines: New Freedoms Through Computer Screens—A Minority Report.* Chicago: Theodor H. Nelson.

Norberg, Arthur L. 1990. "High-Technology Calculation in the Early 20th Century: Punched Card Machinery in Business and Government." *Technology and Culture* 31, no. 4: 753–79.

Norberg, Arthur L., and Judy E. O'Neill. 2000. *Transforming Computer Technology: Information Processing for the Pentagon, 1962–1986.* Baltimore: Johns Hopkins University Press.

Nyce, James M., and Paul Kahn. 1989. "Innovation, Pragmaticism, and Technological Continuity: Vannevar Bush's Memex." *Journal of the American Society for Information Science* 40, no. 3: 214–20.

Nyce, James M., and Paul Kahn, eds. 1991. *From Memex to Hypertext: Vannevar Bush and the Mind's Machine.* Boston: Academic Press.

O'Neill, Judy. 1992. "The Evolution of Interactive Computing Through Time Sharing and Networking." Ph.D. diss., University of Minnesota. Available from University Microfilms International, Ann Arbor, Mich.

Owens, Larry. 1986. "Vannevar Bush and the Differential Analyzer: The Text and Context of an Early Computer." *Technology and Culture* 27, no. 1: 63–95.

Parker Pearson, Jamie, ed. 1992. *Digital at Work: Snapshots from the First Thirty-Five Years.* Burlington, Mass.: Digital Press.

Parker, William N., ed. 1986. *Economic History and the Modern Economist.* Oxford: Basil Blackwell.

Parkhill, D. F. 1966. *The Challenge of the Computer Utility.* Reading, Mass.: Addison-Wesley.

Petre, Peter. 1985. "The Man Who Keeps the Bloom on Lotus" (profile of Mitch Kapor). *Fortune,* 10 June, pp. 92–100.

Phister, Montgomery, Jr. 1979. *Data Processing: Technology and Economics.* Santa Monica, Calif.: Digital Press and Santa Monica Publishing.

Plugge, W. R., and M. N. Perry. 1961. "American Airlines' 'SABRE' Electronic Reservations System." *Proceedings of the AFIPS 1961 Western Joint Computer Conference* (pp. 592–601), where TK, dates TK. Washington D.C.: Spartan Books.

Poole, Steven. 2000. *Trigger Happy: The Inner Life of Videogames.* London: Fourth Estate.

Pugh, Emerson W. 1984. *Memories That Shaped an Industry: Decisions Leading to IBM System/360.* Cambridge, Mass.: MIT Press.

———. 1995. *Building IBM: Shaping an Industry and Its Technology.* Cambridge, Mass.: MIT Press.

Pugh, Emerson W., Lyle R. Johnson, and John H. Palmer. 1991. *IBM's 360 and Early 370 Systems.* Cambridge, Mass.: MIT Press.

Ralston, Anthony, Edwin D. Reilly, and David Hemmendinger, eds. 2000. *Encyclopedia of Computer Science,* 4th ed. London: Nature Publishing.

Randell, Brian. 1982. *Origins of Digital Computers: Selected Papers.* New York: Springer-Verlag.

Redmond, Kent C., and Thomas M. Smith. 1980. *Project Whirlwind: The History of a Pioneer Computer.* Bedford, Mass.: Digital.

———. 2000. *From Whirlwind to MITRE: The R&D Story of the SAGE Air Defense Computer.* Cambridge, Mass.: MIT Press.

Reid, Robert H. 1997. *Architects of the Web: 1,000 Days That Built the Future of Business.* New York: John Wiley.

Richardson, L. F. 1922. *Weather Prediction by Numerical Process.* Cambridge: Cambridge University Press.

Ridenour, Louis N. 1947. *Radar System Engineering.* New York: McGraw-Hill.

Rifkin, Glenn, and George Harrar. 1985. *The Ultimate Entrepreneur: The Story of Ken Olsen and Digital Equipment Corporation.* Chicago: Contemporary Books.

Ritchie, Dennis M. 1984a. "The Evolution of the UNIX Time-Sharing System." *AT&T Bell Laboratories Technical Journal* 63, no. 8: 1577–93.

———. 1984b. "Turing Award Lecture: Reflections on Software Research." *Communications of the ACM* 27, no. 8: 758.

Rodgers, William. 1969. *Think: A Biography of the Watsons and IBM.* New York: Stein & Day.

Rojas, Raul, and Ulf Hashagen, eds. 2000. *The First Computers: History and Architectures.* Cambridge, Mass.: MIT Press.

Roland, Alex, and Philip Shiman. 2002. *Strategic Computing: DARPA and the Quest for Machine Intelligence, 1983–1993.* Cambridge, Mass.: MIT Press.

Rosen, Saul, ed. 1967. *Programming Systems and Languages.* New York: McGraw-Hill.

Salus, Peter H. 1994. *A Quarter Century of Unix.* Reading, Mass.: Addison-Wesley.

Sammet, Jean E. 1969. *Programming Languages: History and Fundamentals.* Englewood Cliffs, N.J.: Prentice-Hall.

Saxenian, AnnaLee. 1994. *Regional Advantage: Culture and Competition in Silicon Valley and Route 128.* Cambridge, Mass.: Harvard University Press.

Schein, Edgar H., Peter S. DeLisi, Paul J. Kampas, and Michael M. Sonduck. 2003. *DEC Is Dead, Long Live DEC: The Lasting Legacy of Digital Equipment Corporation.* San Francisco: Berrett-Koehler.

Scientific American. 1966. *Information.* San Francisco: W. H. Freeman.

———. 1984. *Computer Software.* Special issue. September.

———. 1991. *Communications, Computers and Networks.* Special issue. September.

Sculley, John, and J. A. Byrne. 1987. *Odyssey: Pepsi to Apple . . . A Journey of Adventure, Ideas, and the Future.* New York: Harper & Row.

Sheehan, R. 1956. "Tom Jr.'s I.B.M." *Fortune*, September, p. 112.

Sigel, Efrem. 1984. "The Selling of Software." *Datamation,* 15 April, pp. 125–28.

Slater, Robert. 1987. *Portraits in Silicon.* Cambridge, Mass.: MIT Press.

Smith, Crosbie, and M. Norton Wise. 1989. *Energy and Empire: A Biographical Study of Lord Kelvin.* Cambridge: Cambridge University Press.

Smith, David C. 1986. *H. G. Wells: Desperately Mortal: A Biography.* New Haven, Conn.: Yale University Press.

Smith, Douglas K., and Robert C. Alexander. 1988. *Fumbling the Future: How Xerox Invented, Then Ignored, the First Personal Computer.* New York: Morrow.

Smulyan, Susan. 1994. *Selling Radio: The Commercialization of American Broadcasting, 1920–1934.* Washington, D.C.: Smithsonian Institution Press.

Sobel, Robert. 1981. *IBM: Colossus in Transition.* New York: Times Books.

Spector, Robert. 2000. *Amazon.com: Get Big Fast.* New York: Random House.

Springer, Charles A. 1966. "Data Processing—1890 Style." *Datamation,* July, p. 44.

Stern, Nancy. 1981. *From ENIAC to UNIVAC: An Appraisal of the Eckert-Mauchly Computers.* Bedford, Mass.: Digital Press.

Stross, Randall E. 1996. *The Microsoft Way: The Real Story of How the Company Outsmarts Its Competition.* Reading, Mass.: Addison-Wesley.

Swade, Doron. 1991. *Charles Babbage and His Calculating Engines.* London: Science Museum.

———. 2001. *The Difference Engine: Charles Babbage and the Quest to Build the First Computer.* New York: Viking.

Swisher, Kara. 1998. *AOL.COM: How Steve Case Beat Bill Gates, Nailed the Netheads, and Made Millions in the War for the Web.* New York: Times Books.

Taviss, Irene. 1970. *The Computer Impact.* Englewood Cliffs, N.J.: Prentice-Hall.

Toole, B. A. 1992. *Ada, the Enchantress of Numbers.* Mill Valley, Calif.: Strawberry Press.

Truesdell, Leon E. 1965. *The Development of Punch Card Tabulation in the Bureau of the Census, 1890–1940*. Washington, D.C.: U.S. Department of Commerce.

Turck, J. A. V. 1921. *Origin of Modern Calculating Machines*. Chicago: Western Society of Engineers.

Usselman, Steven W. 1993. "IBM and Its Imitators: Organizational Capabilities and the Emergence of the International Computer Industry." *Business and Economic History* 22, no. 2: 1–35.

Uttal, Bro. 1983. "The Lab That Ran Away from Xerox." *Fortune*, September, pp. 97–102.

Valley, George E., Jr. 1985. "How the SAGE Development Began." *Annals of the History of Computing* 7, no. 3: 196–226.

van den Ende, Jan. 1992. "Tidal Calculations in the Netherlands." *Annals of the History of Computing* 14, no. 3: 23–33.

———. 1994. *The Turn of the Tide: Computerization in Dutch Society, 1900–1965*. Delft: Delft University Press.

Veit, Stan. 1993. *Stan Veit's History of the Personal Computer*. Asheville, N.C.: WorldComm.

Waldrop, M. Mitchell. 2001. *The Dream Machine: J. C. R. Licklider and the Revolution That Made Computing Personal*. New York: Viking.

Wallace, James, and Jim Erickson. 1992. *Hard Drive: Bill Gates and the Making of the Microsoft Empire*. New York: John Wiley.

Wang, An, with Eugene Linden. 1986. *Lessons: An Autobiography*. Reading, Mass.: Addison-Wesley.

Watkins, Ralph. 1984. *A Competitive Assessment of the U.S. Video Game Industry*. Washington, D.C.: U.S. Department of Commerce.

Watson, Thomas, Jr., and Peter Petre. 1990. *Father and Son & Co: My Life at IBM and Beyond*. London: Bantam Press.

Wells, H. G. 1938. *World Brain*. London: Methuen.

———. 1995. *World Brain: H. G. Wells on the Future of World Education*. Ed. A. J. Mayne. 1928. London: Adamantine Press.

Wexelblat, Richard L., ed. 1981. *History of Programming Languages*. New York: Academic Press.

Wildes, K. L., and N. A. Lindgren. 1986. *A Century of Electrical Engineering and Computer Science at MIT, 1882–1982*. Cambridge, Mass.: MIT Press.

Wilkes, Maurice V. 1985. *Memoirs of a Computer Pioneer*. Cambridge, Mass.: MIT Press.

Wilkes, Maurice V., David J. Wheeler, and Stanley Gill. 1951. *The Preparation of Programs for an Electronic Digital Computer*. Reading, Mass.: Addison-Wesley. Reprint, with an introduction by Martin Campbell-Kelly, Los Angeles: Tomash Publishers, 1982. See also volume 1, Charles Babbage Institute Reprint Series for the History of Computing, introduction by Martin Campbell-Kelly.

Williams, Frederik C. 1975. "Early Computers at Manchester University." *The Radio and Electronic Engineer* 45, no. 7: 327–31.

Williams, Michael R. 1997. *A History of Computing Technology.* Englewood Cliffs, N.J.: Prentice-Hall.

Wise, Thomas A. 1966a. "I.B.M.'s $5,000,000,000 Gamble." *Fortune,* September, p. 118.

———. 1966b. "The Rocky Road to the Market Place." *Fortune,* October, p. 138.

Yates, JoAnne. 1982. "From Press Book and Pigeonhole to Vertical Filing: Revolution in Storage and Access Systems for Correspondence." *Journal of Business Communication* 19 (Summer): 5–26.

———. 1989. *Control Through Communication: The Rise of System in American Management.* Baltimore, Md.: Johns Hopkins University Press.

———. 1993. "Co-evolution of Information-Processing Technology and Use: Interaction Between the Life Insurance and Tabulating Industries." *Business History Review* 67, no. 1: 1–51.

Yates, JoAnne, and John Van Maanen. 2001. *Information Technology and Organizational Transformation: History, Rhetoric, and Preface.* Thousand Oaks, Calif.: Sage.

Zachary, G. Pascal. 1997. *Endless Frontier: Vannevar Bush, Engineer of the American Century.* New York: Free Press.

:: INDEX